CISM International Centre for Mechanical Sciences

Courses and Lectures

Volume 2

For more than 40 years the book series edited by CISM, "International Centre for Mechanical Sciences: Courses and Lectures", has presented groundbreaking developments in mechanics and computational engineering methods. It covers such fields as solid and fluid mechanics, mechanics of materials, micro- and nanomechanics, biomechanics, and mechatronics. The papers are written by international authorities in the field. The books are at graduate level but may include some introductory material.

Fernando Fraternali · Julian J. Rimoli
Editors

Tensegrity Systems

Basic Concepts, Mechanical Metamaterials, Biotensegrity

Springer

Editors
Fernando Fraternali
Department of Civil Engineering
University of Salerno
Fisciano, Italy

Julian J. Rimoli
Department of Mechanical and Aerospace Engineering
University of California Irvine
Irvine, CA, USA

ISSN 0254-1971 ISSN 2309-3706 (electronic)
CISM International Centre for Mechanical Sciences
ISBN 978-3-031-82282-7 ISBN 978-3-031-82283-4 (eBook)
https://doi.org/10.1007/978-3-031-82283-4

This Springer imprint is published by the registered company Springer Nature Switzerland AG
The registered company address is: Gewerbestrasse 11, 6330 Cham, Switzerland

Preface

The first definitions of tensegrity systems were introduced by the American architect Richard Buckminster Fuller in the late 1950s and early 1960s. Inspired by the remarkable sculptures of artist Kenneth Snelson, a student of his at Black Mountain College (North Carolina) in 1949, Fuller initially described tensegrity as 'a system of construction which utilizes the tensile properties of structural materials to the fullest advantage'. Later, he refined the concept, defining tensegrity as 'a contraction of tension integrity, a structure whose shape is maintained by the tensional behaviors of the system, rather than by compressional behaviors' (Synergetics Dictionary, October 1963). Kenneth Snelson referred to his sculptures as 'endoskeletal, floating compression structures' to highlight that compressive members are isolated and contained within a network of tensile members. These definitions aim to describe systems commonly referred to as 'pure tensegrities', 'classical tensegrities', or 'floating-compression tensegrity systems', in which compressive members, isolated from direct contact with one another, are interconnected by pre-stretched tensile elements. Such structures typically display infinitesimal mechanisms, stabilized by an internal, self-equilibrated state of prestress, induced through the pre-tensioning of tensile members.

Over the years, the term tensegrity has evolved and can now describe a broader range of assemblies: 'strings' or 'cables' that carry only tensile forces, 'struts' that carry only compressive forces, and 'bars' that can carry either compressive or tensile forces. This expanded definition includes systems with infinitesimal mechanisms and self-equilibrated force systems (prestressable systems), as well as those lacking these properties. It does not require compressive members to be isolated; indeed, one can design tensegrities with any number of compressive elements converging at a single node. Additionally, flexible tensegrity systems may refer to generalized tensegrity structures where the bars are allowed to buckle freely. Interest in these structures is often inspired by nature, as demonstrated in several examples examined in detail throughout this book.

The book presents topics from an advanced course held at the International Center for Mechanical Sciences in Udine, Italy, in September 2024. It aims to illustrate the unique mechanical behaviors of tensegrity systems and their applications in mechanical metamaterials, space structures, and biomechanical models. Several studies have shown that by controlling the mechanical response of tensegrity structures through internal and external prestress, it is possible to adjust the speed of mechanical waves within these systems, creating tunable bandgap structures. Furthermore, the geometrically nonlinear response exhibited by several tensegrity systems allows for the support of either compression or rarefaction solitary wave dynamics. These behaviors can be effectively utilized to design novel devices capable of focusing mechanical waves in narrow regions of space, as well as innovative impact protection systems.

The book begins with an introduction to the basic concepts and calculation methods for tensegrity systems and their minimal-mass design (Chaps. 1 and 2). These sections provide an analytical treatment of the statics and dynamics of tensegrity systems, covering the fundamentals of compatibility, equilibrium, optimal design, stability, and equations of motion. Chapters 3 and 4 explore the metamaterial behaviors of tensegrity systems associated with bandgap and solitary wave dynamics. Chapter 5 presents a mechanical model of flexible tensegrities, illustrating how harnessing the buckling of bars in such systems can result in structures with exceptional energy absorption capabilities, suitable for applications such as planetary landers or lattice metamaterials. The section on metamaterial applications concludes with Chap. 6, which discusses the extreme mechanical behaviors achievable in tensegrity-inspired lattice structures exhibiting both soft and stiff deformation modes. Chapters 7 and 8 address the multifaceted field of biotensegrity. These sections demonstrate that tensegrity structures, which mimic the mechanics of biological systems— a concept known as biotensegrity—can replicate key features of living systems across multiple scales, including their adaptive, shape changing strategies. The book concludes with Chap. 9, providing an overview of current rapid prototyping techniques for tensegrity systems, along with a discussion of open questions and future research opportunities in this field.

Fisciano, Italy Fernando Fraternali
Irvine, USA Julian J. Rimoli

Contents

Chapter 1
Basic Tensegrity Concepts and Calculation Methods

Andrea Micheletti

Abstract In this chapter, the fundamental notions about tensegrity systems are presented, together with the basic calculation methods. After giving a general introduction to the main tensegrity concepts, we dive into the analytical treatment of the statics and dynamics of tensegrity systems. In particular, we introduce the equilibrium and kinematic-compatibility equations and describe the four structural types to which tensegrity systems may belong, characterized by the presence or absence of selfstress states and internal mechanisms. We then pass to quasistatic processes and give the notions of prestress stability and superstability, which depend on the properties of the geometric stiffness and tangent stiffness operators. Notable examples are then reviewed to illustrate the nonlinear behavior of tensegrity systems. Dynamic processes are described next by deriving the motion equation, both for nonlinear large-displacement processes and, in its linearized form, for small-amplitude motions about a certain equilibrium configuration. The last part of the chapter introduces form-finding methods, and presents in details three of them with simple implementation: an analytical method for symmetric systems, the virtual cocoon method, and a marching algorithm to change the shape of a tensegrity system by controlling edge lengths. The latter method is then applied to the case study of a deployable ring for space antennas.

Keywords Tensile integrity · Balloon analogy · Selfstress states · Infinitesimal mechanisms · Singular-value decomposition · Geometric stiffness · Prestress stability · Superstability · Bistability · Form finding · Deployable structures.

A. Micheletti (✉)
Dipartimento di Ingegneria Civile e Ingegneria Informatica, University of Rome Tor Vergata, Via Politecnico 1, 00133 Rome, Italy
e-mail: micheletti@ing.uniroma2.it

F. Fraternali and J. J. Rimoli (eds.), *Tensegrity Systems*, CISM International Centre for Mechanical Sciences 2, https://doi.org/10.1007/978-3-031-82283-4_1

1.1 Introduction

In this section, we expound on the main basic notions and concepts related to tensegrity structures. We refer the reader to the review articles (Sultan, 2009; Micheletti & Podio-Guidugli, 2022; Gómez-Jáuregui et al., 2023) and the textbooks (Skelton et al., 2002; Motro, 2003; Skelton & de Oliveira, 2009; Connelly & Guest, 2022) for a more detailed introduction to tensegrity systems and extended lists of references.

We begin with a concise timeline on tensegrities. At the end of the 1940s and during the 1950s, we have the first studies by Kenneth Snelson, Richard Buckminster Fuller, in the United States, and by George D. Emmerich in France, who started popularizing the tensegrity concept and got their respective patents granted in the 1960s. A detailed report on the origins of tensegrities is given in Gómez-Jáuregui (2009). Starting in the late 1970s, we find the first rigorous studies in engineering and mathematics (Calladine, 1978; Roth & Whiteley, 1981; Connelly, 1982; Motro, 1984; Pellegrino & Calladine, 1986; Pellegrino, 1990; Motro, 1992; Connelly & Whiteley, 1996), while in 1988 the first large-span roof inspired by one of Buckminster Fuller's designs, the Geiger dome, was realized in Korea (Geiger et al., 1986). In 1997, Robert E. Skelton patented the first cable-driven deployable tensegrity system (Skelton, 1997). The concept of tensegrity was also proposed in biology by (Ingber, 1998), and more recently it was put forward in biomechanics (Levin et al., 2017; Oh et al., 2022). From 2005, we find the first studies in robotics by Hod Lipson and coworkers (Paul et al., 2005), and various tensegrity robots for space applications were later proposed (Bruce et al., 2014; Kim et al., 2017). More recently, tensegrity metamaterials have been considered by Fernando Fraternali and coworkers (Fraternali et al., 2012, 2014b; Amendola et al., 2018), followed by Rimoli and Pal (2017) and Gilewski and AlSabouni-Zawadzka (2020). In the same years, the challenging task of rapid prototyping was tackled (Amendola et al., 2015; Liu et al., 2017; Lee et al., 2020).

1.1.1 Tensegrity Definitions

Different definitions of tensegrity systems can be found in the literature. When looking at one of Kenneth Snelson's great sculptures, one can recognize a prestressed assembly of cables and bars connected to each other at their ends, with each bar connected only to cables. The assembly is simply supported on the ground, that is, there is no need of external anchorages. Systems of this type are often referred to as *pure tensegrities*, *classical tensegrities*, or *floating-compression tensegrities*.

It is convenient to adopt the definition given by mathematicians in rigidity theory (cf. Connelly & Whiteley, 1996), who consider tensegrity systems as assemblies of bars, cables, and struts connected at their endpoints, with no further specification; a definition aimed at devising the most general theoretical framework. More specifically, bars can carry both tension and compression, cables can carry tension only, and struts can carry compression only. This general definition encompasses systems that feature infinitesimal mechanisms and self-equilibrated force systems (prestressable systems), as well as those that do not exhibit such properties.

In this chapter, we consider assemblies of cables and bars and assume that the cables are in tension and the bars are in compression, if not specified otherwise. We will indicate, when necessary, whether the tensegrity system under consideration is pure or not.

1.1.2 Form-Finding Property and Form-Finding Problems

Many of the tensegrity sculptures realized by Snelson possess the *form-finding property*, which can be described as follows. Let us consider a tensegrity structure composed of a certain number of cables and bars, and let us assume that all the connections between such elements are realized except for those needed to assemble the last element. The partial assembly at this stage has significant flexibility and no stiffness. The system will acquire its shape and become stiff only when the distance between the two points to be connected will reach a minimum if the last element is a cable or a maximum if the last element is a bar in compression. By forcing the two points to get closer to each other, in the former case, or further apart, in the latter case, and then connecting the last element, the assembly will acquire a prestress state involving all its elements. In addition, the stiffness of the resulting structure will depend on the amount of prestressing imparted by performing this operation. Figure 1.1 exemplifies the form-finding property for a simple two-element system.

We speak of the *form-finding problem* when we want to determine in advance the stable load-free equilibrium configuration of a prestressed structure. The unknowns may include nodal positions, the node-element incidence relation, and the cable/bar nature of the elements. A variety of methods have been devised in order to solve

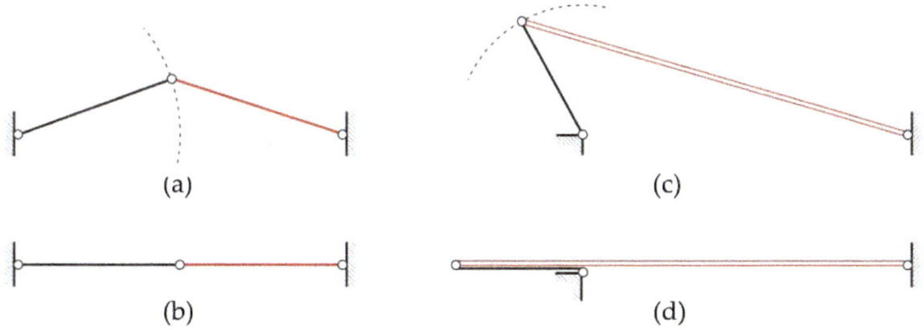

Fig. 1.1 The form-finding property for a two-element system. **a** The length of the element on the left is fixed, the length of the element on the right is not; the central node can only be on the dashed arc shown. **b** On shortening the right element, the collinear configuration is reached; further shortening induces a tensional self-stress state in both elements. **c, d** When the right element is lengthened, in the collinear configuration the right element is in compression, the left element is in tension

form-finding problems for tensegrity structures. This topic will be discussed in the last part of this chapter.

1.1.3 Non-generic Configurations and Minimality

In many cases, the form-finding property is related to a peculiar characteristic of the tensegrity systems to which it applies. The characteristic is that they are minimal systems, because they are composed of a number of elements lower than the number of elements usually required to obtain a rigid structure (see, e.g., Kelly et al., 2019 for a precise definition of rigidity).

It can be seen that a minimal configuration is determined by the solution of an algebraic equation in the coordinates of the elements' endpoints. In this sense, such configurations are referred to as *non-generic*, which means 'special', in contrast to *generic configurations* that do not satisfy that equation. We will see later that the algebraic equation descends from the condition for which the rank of the equilibrium operator of the system is not maximal.

Consider the three-element system in two dimensions shown in Fig. 1.2a, which constitutes a finite mechanism with one degree of freedom. By shortening one of the three elements, the configuration of Fig. 1.2b in which the four nodes are aligned is reached, and the system acquires a state of selfstress with the three elements in tension. If elastic bars are assembled in this configuration, the system can carry external loads with limited deformation. The aligned configuration is non-generic, whereas when the nodes are not aligned, the configuration is generic. When the nodes are not aligned, one more element would be necessary to realize a rigid structure. In the aligned configuration, three elements are enough to make it rigid. We define the *minimality ratio* d_m as the number of missing elements divided by the number of elements necessary to realize a rigid structure in a generic configuration. For the just-described three-element system, $d_m = 1/4$. Analogous considerations apply if such a system is embedded in three dimensions, with $d_m = 3/6$.

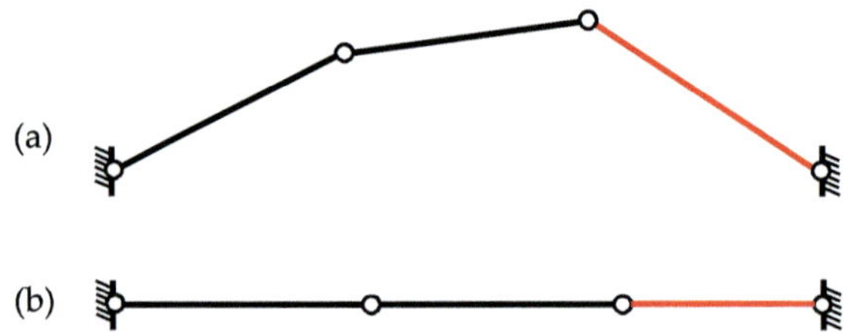

Fig. 1.2 **a** A three element system in a generic configuration. One more element would be necessary to make it rigid. **b** The non-generic configuration reached on shortening one element acquires one selfstress state with the elements in tension and becomes rigid

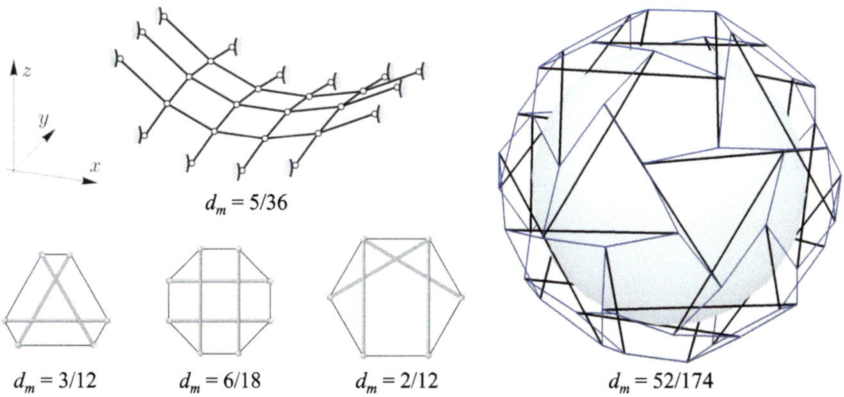

Fig. 1.3 Examples of minimal systems that are also superstable, with indicated the corresponding minimality ratio. A sphere is drawn inside the system on the right for ease of visualization

Figure 1.3 shows some examples of minimal systems in three dimensions together with their minimality ratio: a cable net, three cable-bar polygons, and a tensegrity balloon.

1.1.4 Balloon Analogy and Superstability

Snelson referred to his realizations as *endoskeletal systems* to remark the fact that the rigid elements, the bars, are always enveloped by the network of flexible elements, the cables. In relation to this fact, Buckminster Fuller proposed the *balloon analogy* (Motro, 2003): the network of cables and the bars of a tensegrity system are analogous, respectively, to the skin and the air under pressure of an inflated balloon. This analogy also reflects the conditions that a tensegrity system with no prestress has no stiffness, just as a deflated balloon, and that stiffness increases by increasing prestress.

A special property of some tensegrity systems is that of superstability, precisely defined later, which indicates that the system is stable independently of the prestress level and material properties. Stable tensegrities that do not have this property can become unstable when their prestress is increased or when their elements are realized with softer materials. It can be observed that the balloon analogy also applies to many superstable tensegrities. All the examples shown in Fig. 1.3 are superstable systems.

1.1.5 Tensegrity Design: Advantages and Challenges

Here we describe the principal features that make tensegrities interesting for applications, together with some of the challenges that should be overcome for successful realizations.

One of the most important advantages of tensegrity systems is that, due to the presence of cables, they can be easily folded/deployed, and, in general, they can be employed as variable-geometry cable-driven structures (see, e.g, Sultan & Skelton, 1998, 2003; Rhode-Barbarigos et al., 2012; Bruce et al., 2014; Micheletti et al., 2018; Ma et al., 2022). In addition, pure tensegrities have a reduced risk of failure as deployable structures because they do not have mechanical joints between the rigid elements, eliminating friction-related problems (Ganga et al., 2016).

The other relevant aspect is their complex nonlinear response, both in statics and in dynamics, which can be exploited in diverse applications. In fact, it has been observed that tensegrities can have a marked stiffening and softening behavior (Oppenheim & Williams, 2000; Fraternali et al., 2014a), different regimes of bistability and multistability (Xu & Luo, 2010; Micheletti, 2013; Micheletti et al., 2023), and they can support the propagation of solitary waves (Fraternali et al., 2012, 2014b; Davini et al., 2016; Micheletti et al., 2019). Moreover, such nonlinear response can be tuned in different ways, for example, by varying the prestress, or by slightly modifying the geometry (Amendola et al., 2018; Micheletti et al., 2023). These qualities can be exploited to realize novel structures and tensegrity-based metamaterials with unprecedented properties.

The complex nonlinear behavior of tensegrity systems constitutes one of the main challenges for designers. There are different levels of design for these peculiar structures. Typically, a form finding problem needs to be solved in order to find a stable equilibrium configuration. At the same time, such configuration is associated with certain selfstress states, in which cables should be sufficiently in tension to avoid slackening under external loads, while bars should not be excessively compressed to avoid local buckling. In addition, global stability must also be ensured. For a deployable tensegrity structure, these requirements should be fulfilled for a family of configurations, ranging from the folded state to the deployed state. In case the design objective is a tailored nonlinear response, the task is even more complex. Furthermore, these aspects are specific to the node-edge incidence relationship determined at the form-finding stage, which makes the selection of an appropriate tensegrity system a crucial step for a successful design.

Another main challenge of tensegrity structures is their realization. The fact that a tensegrity assembly is not stiff until the last elements are in place and a prestress state is established makes it necessary to have temporary supports during the construction process. For this reason, there is no established method for rapid prototyping these structures, and conventional additive manufacturing techniques cannot be applied to tensegrity systems, making manual assembly the most popular construction method.

Finally, we need to mention that tensegrity systems have been often referred to as lightweight structures. However, some authors have shown that this is not the case for pure tensegrities (Wang, 2004; Tibert, 2008; Hanaor, 2012; Cadoni and Micheletti, 2012; Feron et al., 2019), as Snelson himself acknowledged. There are essentially two reasons for this. First, since pure tensegrities have a small number of bars and a large number of cables, on average bars are longer and carry larger axial forces Cadoni and Micheletti, 2012, a fact that makes their design against buckling taxing. Second, having a fixed tension-compression pattern with discontinuous compressive elements, internal forces cannot realize a continuous compression path in response to external loads, as happens in conventional structures subjected to bending (Tibert, 2008), and the selfstress level must be large enough to avoid loss of tension in cables. For these reasons, tensegrities that are not pure, that is, with contiguous bars and/or external anchorages, have better chances of being able to realize lightweight structures. For example, this is the case of many realized tensegrity domes, which are anchored to a peripheral compression ring.

1.2 Employed Notation

This chapter uses the following symbols for key variables: number of nodes (n_n), number of edges (n_e), position vector of node i ($\boldsymbol{p}_i$), vector of all nodal positions ($\boldsymbol{p}$), length of edge ij (l_{ij}), vector of all edge lengths ($\boldsymbol{l}$), axial force of edge ij (t_{ij}), vector of all axial forces ($\boldsymbol{t}$), force density of edge ij (q_{ij}), vector of all force densities ($\boldsymbol{q}$), diagonal operator containing the force densities ($\boldsymbol{Q}$), unit vector pointing from node i to node j ($\boldsymbol{n}_{ij}$), load vector on node i ($\boldsymbol{f}_i$), vector of all nodal loads ($\boldsymbol{f}$), equilibrium operator in terms of axial forces ($\boldsymbol{A}$), column of $\boldsymbol{A}$ corresponding to edge ij ($\boldsymbol{a}_{ij}$), equilibrium operator in terms of force densities ($\widehat{\boldsymbol{A}}$), column of $\widehat{\boldsymbol{A}}$ corresponding to edge ij ($\widehat{\boldsymbol{a}}_{ij}$), velocity vector of node i ($\boldsymbol{v}_i$), vector of all nodal velocities ($\boldsymbol{v}$), displacement vector of node i ($\boldsymbol{u}_i$), vector of all nodal displacements ($\boldsymbol{u}$), elongation of edge ij (e_{ij}), vector of all edges elongations ($\boldsymbol{e}$), kinematic-compatibility operator ($\boldsymbol{B}$), elongation coefficient of edge ij ($\widehat{e}_{ij}$), vector of all elongation coefficients ($\widehat{\boldsymbol{e}}$), kinematic-compatibility operator in terms of elongation coefficients ($\widehat{\boldsymbol{B}}$), rank of $\boldsymbol{A}$ (r), number of independent selfstress states (n_s), number of independent mechanisms (n_m), stiffness constant of edge ij (k_{ij}), diagonal operator containing the stiffness constants ($\boldsymbol{D}$), total energy (E_{tot}), elastic potential energy (E_{el}), potential energy of the loads (E_{f}), tangent stiffness operator ($\boldsymbol{K}_T$), connectivity operator of edge ij ($\boldsymbol{C}_{ij}$), stress matrix ($\boldsymbol{S}$), material stiffness operator ($\boldsymbol{K}_M$), geometric stiffness operator ($\boldsymbol{K}_G$), geometric load vector associated to mechanism $\boldsymbol{u}_m$ ($\boldsymbol{g}_m$), energy dissipation ($\mathfrak{D}$), damping constant of node i ($d_i^{(p)}$), nodal damping operator ($\boldsymbol{D}^{(p)}$), damping constant of edge ij ($d_{ij}^{(l)}$), edge damping operator ($\boldsymbol{D}^{(l)}$), kinetic energy (E_{kin}), mass operator ($\boldsymbol{M}$), damping operator ($\boldsymbol{C}_d$).

1.3 Structural Classification of Bar Frameworks

Studying the properties of bar frameworks is essential for understanding those of tensegrity systems. Here we review the linear theory of bar frameworks to familiarize ourselves with the relevant definitions and equations and to categorize structures into four classes. We will then see how tensegrity systems can fit these classes after replacing bars in tension with cables.

A pin-connected bar framework can be described in terms of three sets, $\mathcal{F} = \{\mathcal{N}, \mathcal{E}, \mathcal{P}\}$, a set of vertices, or nodes, $\mathcal{N} = \{i = 1, \ldots, n_n\}$, a set of edges connecting node pairs $\mathcal{E} = \{ij : i, j \in \mathcal{N}, i \neq j\}$, with n_e is the number of edges, and a set of points in the Euclidean space E associated to the nodes, $\mathcal{P} = \{P_i \in E, i = 1, \ldots, n_n\}$, which determine the *configuration* of the framework. We say that $\mathcal{G} = \{\mathcal{N}, \mathcal{E}\}$ is the underlying graph of the framework. By choosing a reference frame $\{O; x, y, z\}$ for E, a configuration is expressed by the $(3n_n)$-dimensional vector $\boldsymbol{p}$ containing all the position vectors $\boldsymbol{p}_i$ of the points P_i, $i = 1, \ldots, n_n$,

$$[\boldsymbol{p}] = \begin{bmatrix} \vdots \\ \boldsymbol{p}_i \\ \vdots \end{bmatrix} = \begin{bmatrix} \vdots \\ x_i \\ y_i \\ z_i \\ \vdots \end{bmatrix} .$$

An edge ij connecting nodes i and j has length $l_{ij} = |\boldsymbol{p}_j - \boldsymbol{p}_i|$, and it carries the axial force t_{ij}, positive if tensile. The unit vector pointing from P_i to P_j is $\boldsymbol{n}_{ij} = (\boldsymbol{p}_j - \boldsymbol{p}_i)/l_{ij}$, with $\boldsymbol{n}_{ji} = -\boldsymbol{n}_{ij}$.

1.3.1 Equilibrium

Pin connections are assumed to be frictionless and the loads are point forces applied to nodes only; thus, bars are not subjected to shear, bending or twisting. We assume that a certain number of nodes are constrained to not move in any direction. The equilibrium equation of an unconstrained node i subjected to external and internal forces is written as

$$\boldsymbol{f}_i + \sum_{\substack{j \in \mathcal{N}: \\ ij \in \mathcal{E}}} t_{ij}\, \boldsymbol{n}_{ij} = \boldsymbol{0} ,$$

or, equivalently,

$$\sum_j \frac{t_{ij}}{l_{ij}} (\boldsymbol{p}_i - \boldsymbol{p}_j) = \boldsymbol{f}_i , \tag{1.1}$$

where the summation is extended to every nodes j connected to node i by an edge. By defining the force density, or force coefficient, of edge ij as $q_{ij} = t_{ij}/l_{ij}$ the equilibrium equation of the i-th node becomes

$$\sum_j q_{ij}\,(\boldsymbol{p}_i - \boldsymbol{p}_j) = \boldsymbol{f}_i\,. \tag{1.2}$$

The equilibrium equations of all nodes can be written in compact form as

$$\boldsymbol{A}\,\boldsymbol{t} = \boldsymbol{f}\,, \tag{1.3}$$

or, in terms of force densities,

$$\widehat{\boldsymbol{A}}\,\boldsymbol{q} = \boldsymbol{f}\,,$$

where $\boldsymbol{A}$ and $\widehat{\boldsymbol{A}}$ are the linear operator expressed by (1.1) and (1.2), respectively, written for each node, $\boldsymbol{t},\,\boldsymbol{q} \in \mathcal{V}^{n_e}$ are the n_e-dimensional vectors containing the axial forces t_{ij} and force densities q_{ij}, respectively, of all edges, and $\boldsymbol{f} \in \mathcal{V}^{3n_n}$ is the $(3n_n)$-dimensional vector containing all nodal loads $\boldsymbol{f}_i, i = 1, \ldots, n_n$. The advantage of writing the equilibrium equations in terms of force densities is that the operator $\widehat{\boldsymbol{A}}$ depends linearly on the nodal coordinates, which is a useful property for form-finding problems.

The *selfstress states* are defined as the axial forces $\boldsymbol{t}_s$ (force densities $\boldsymbol{q}_s$) in equilibrium with null loads, solutions of the system of equation $\boldsymbol{A}\boldsymbol{t}_s = \boldsymbol{0}$ ($\widehat{\boldsymbol{A}}\boldsymbol{q}_s = \boldsymbol{0}$), i.e., the elements of the nullspace of $\boldsymbol{A}$ ($\widehat{\boldsymbol{A}}$). The dimension n_s of this nullspace corresponds to the degree of static indeterminacy, that is, the number of independent selfstress states of the framework. It is worth remarking that, since $\widehat{\boldsymbol{A}} = \boldsymbol{A}\,\mathrm{diag}(\boldsymbol{l})$, with $\boldsymbol{l}$ the vector containing all edge lengths, the two operators $\boldsymbol{A}$ and $\widehat{\boldsymbol{A}}$ have the same rank.

1.3.2 Kinematic Compatibility

A motion of the framework is a smooth one-parameter family of configurations $\boldsymbol{p}(t)$. By taking the time derivative of the squared length of an edge ij, $l_{ij}^2 = (\boldsymbol{p}_i - \boldsymbol{p}_j)\cdot(\boldsymbol{p}_i - \boldsymbol{p}_j)$, we have

$$(l_{ij}^2)^{\boldsymbol{\cdot}} = 2\,l_{ij}\dot{l}_{ij} = 2\,(\boldsymbol{p}_i - \boldsymbol{p}_j)\cdot(\boldsymbol{v}_i - \boldsymbol{v}_j)\,,$$

where $\boldsymbol{v}_i = \dot{\boldsymbol{p}}_i$. Therefore,

$$\frac{\boldsymbol{p}_i - \boldsymbol{p}_j}{l_{ij}}\cdot(\boldsymbol{v}_i - \boldsymbol{v}_j) = \boldsymbol{n}_{ji}\cdot(\boldsymbol{v}_i - \boldsymbol{v}_j) = \dot{l}_{ij}\,, \tag{1.4}$$

which is the kinematic-compatibility equation relating the lengthening rate $\dot{l}_{ij}$ of edge ij to the nodal velocities of its endpoints, $\boldsymbol{v}_i$, $\boldsymbol{v}_j$.

Given that $\boldsymbol{v}_i = d\boldsymbol{p}_i/dt$, considering a regime of small displacements, the nodal velocity $\boldsymbol{v}_i$ is parallel to the small nodal displacement $\Delta\boldsymbol{p}_i =: \boldsymbol{u}_i$, and we can write the kinematic-compatibility equation analogous to (1.4) relating the elongation $\Delta l_{ij} =: e_{ij}$ of the edge ij to the nodal displacements $\boldsymbol{u}_i$ and $\boldsymbol{u}_j$:

$$\boldsymbol{n}_{ji} \cdot (\boldsymbol{u}_i - \boldsymbol{u}_j) = e_{ij} \,. \tag{1.5}$$

The kinematic-compatibility equations of the whole framework are written as

$$\boldsymbol{B}\,\boldsymbol{v} = \dot{\boldsymbol{e}}\,, \qquad \boldsymbol{B}\,\boldsymbol{u} = \boldsymbol{e}\,, \tag{1.6}$$

where $\boldsymbol{B}$ is the linear operator expressed by (1.4), or (1.5), written for all edges, $\boldsymbol{v}, \boldsymbol{u} \in \mathcal{V}^{3n_n}$ are the $(3n_n)$-dimensional vectors containing the nodal velocities $\boldsymbol{v}_i$ and small displacements $\boldsymbol{u}_i$, respectively, of all nodes, and $\dot{\boldsymbol{e}}, \boldsymbol{e} \in \mathcal{V}^{n_e}$ are the n_e-dimensional vectors containing the lengthening rates $\dot{e}_{ij}$ and small elongations e_{ij}, respectively, of all edges. It is worth remarking that (1.4) and $(1.6)_1$ are exact equations, while (1.5) and $(1.6)_2$ hold true only for first-order changes in length.

The *mechanisms* are defined as the nodal velocities $\boldsymbol{v}_m$ (small displacements $\boldsymbol{u}_m$) that cause null elongation rates (null elongations) of the edges, solutions of $\boldsymbol{B}\boldsymbol{v}_m = \boldsymbol{0}$ ($\boldsymbol{B}\boldsymbol{u}_m = \boldsymbol{0}$), i.e., the elements of the nullspace of $\boldsymbol{B}$. The dimension n_m of this nullspace corresponds to the degree of kinematic indeterminacy, that is, the number of independent mechanisms of the framework.

The power spent by the nodal load $\boldsymbol{f}_i$ on a virtual nodal velocity $\boldsymbol{v}_i$ is $\boldsymbol{f}_i \cdot \boldsymbol{v}_i$, while the power spent by the axial force t_{ij} on the nodal displacements $\boldsymbol{v}_i$, $\boldsymbol{v}_j$ can be computed as

$$t_{ij}\,\boldsymbol{n}_{ij} \cdot \boldsymbol{v}_j + t_{ij}\,\boldsymbol{n}_{ji} \cdot \boldsymbol{v}_i = t_{ij}\,\boldsymbol{n}_{ji} \cdot (\boldsymbol{v}_i - \boldsymbol{v}_j) = t_{ij}\,\dot{e}_{ij}\,,$$

in which the last equality follows from (1.4). By considering the same reference frame and the same numbering of nodes and edges to write (1.1) for all nodes and (1.4) for all edges, it is easy to see that

$$\boldsymbol{B} = \boldsymbol{A}^T\,,$$

and that the virtual power (work) relation holds $\boldsymbol{f} \cdot \boldsymbol{v} = \boldsymbol{t} \cdot \dot{\boldsymbol{e}}$ ($\boldsymbol{f} \cdot \boldsymbol{u} = \boldsymbol{t} \cdot \boldsymbol{e}$), showing that the equilibrium and the kinematic-compatibility operators, $\boldsymbol{A}$ and $\boldsymbol{B}$, contain the same information.

We remark that $\widehat{\boldsymbol{B}} = \widehat{\boldsymbol{A}}^T$, the transpose of the equilibrium operator written with respect to force densities, can be used to rewrite the kinematic compatibility equations $(1.6)_1$ in the form

$$\widehat{\boldsymbol{B}}\boldsymbol{v} = \widehat{\boldsymbol{e}}\,,$$

with $\widehat{e}$ the vector containing the quantities $\widehat{e}_{ij} = l_{ij}\dot{l}_{ij}$, so that the kinematic-compatibility equation for individual edges is written as

$$(\boldsymbol{p}_i - \boldsymbol{p}_j) \cdot (\boldsymbol{v}_i - \boldsymbol{v}_j) = \widehat{e}_{ij} \,. \tag{1.7}$$

1.3.3 Fundamental Subspaces and Structural Classification

On denoting by $\boldsymbol{a}_{ij}$ the column vectors of $\boldsymbol{A}$ corresponding to edge ij, the equilibrium equation (1.3) can be rewritten as

$$\sum_{ij \in \mathcal{E}} \boldsymbol{a}_{ij}\, t_{ij} = \boldsymbol{f} \,,$$

from which we see that admissible loads, the loads that are in equilibrium with internal axial forces, belongs to the column space, or image, of $\boldsymbol{A}$. Given that $\boldsymbol{B} = \boldsymbol{A}^T$, the kinematic-compatibility equation (1.5) for each edge ij can be written as

$$\boldsymbol{a}_{ij} \cdot \boldsymbol{u} = e_{ij} \,, \tag{1.8}$$

which means that the internal mechanisms $\boldsymbol{u}_m$, solutions of $\boldsymbol{A}^T \boldsymbol{u}_m = \boldsymbol{0}$, are orthogonal to the column space of $\boldsymbol{A}$. Thus, admissible loads satisfy the condition

$$\boldsymbol{f} \cdot \boldsymbol{u}_m = 0 \,, \quad \boldsymbol{u}_m \in \text{null}(\boldsymbol{A}^T) \,,$$

which means that they do null work on the mechanisms. The nullspace of $\boldsymbol{A}^T$ and the column space of $\boldsymbol{A}$ are orthogonal complements in $\mathcal{V}^{3n_n}$,

$$\text{col}(\boldsymbol{A}) \perp \text{null}(\boldsymbol{A}^T) \,, \quad \mathcal{V}^{3n_n} = \text{col}(\boldsymbol{A}) \oplus \text{null}(\boldsymbol{A}^T) \,,$$

and, on denoting by r the rank of $\boldsymbol{A}$, equal to the dimension of its column space, we have that

$$3n_n = r + n_m \,. \tag{1.9}$$

Analogously, it can be seen that the nullspace of $\boldsymbol{A}$ and the column space of $\boldsymbol{A}^T$ are orthogonal complements in $\mathcal{V}^{n_e}$,

$$\text{col}(\boldsymbol{A}^T) \perp \text{null}(\boldsymbol{A}) \,, \quad \mathcal{V}^{n_e} = \text{col}(\boldsymbol{A}^T) \oplus \text{null}(\boldsymbol{A}) \,, \quad n_e = r + m \,, \tag{1.10}$$

which means that the work done by selfstress states $\boldsymbol{t}_s$ on kinematically compatible elongations $\boldsymbol{e}$ is always null:

$$\boldsymbol{t}_s \cdot \boldsymbol{e} = 0 \,, \quad \boldsymbol{t}_s \in \text{null}(\boldsymbol{A}) \,.$$

By eliminating r from relations (1.9) and $(1.10)_3$, the so-called *Extended Maxwell's rule* (Calladine, 1978) is obtained:

$$3\,n_n - n_e = n_m - n_s\,. \tag{1.11}$$

It is worth noticing that in writing (1.11) we assumed that constrained nodes are in sufficient number to remove all rigid motions of the framework. In case of a free-standing framework, i.e., all nodes are free nodes, the extended Maxwell's rule can be written as

$$3\,n_n - n_e - 6 = n_m - n_s\,, \tag{1.12}$$

where the number 6 indicates the number of independent rigid-body motions in three dimensions, and n_m indicates the number of internal mechanisms only, that is, those mechanisms corresponding to actual changes of shape of the framework.

Bar frameworks can be given the following classification (see, e.g., Pellegrino, 1990) according to the dimension of the nullspaces of $\boldsymbol{A}$ and $\boldsymbol{A}^T$. It is worth noticing that systems in one class will stay in the same class under projective transformations, those that preserve incidence relations of points, lines, and planes.

Type 1: isostatic frameworks. In this case, we have $n_m = n_s = 0$ and $r = 3\,n_n = n_e$. The framework is statically and kinematically determined, Eqs. (1.3) and (1.6) always have a unique solution for any choice of $\boldsymbol{f}$ and $\boldsymbol{e}$, respectively, and the equilibrium operator $\boldsymbol{A}$ is represented by a square matrix of full rank.

Type 2: underconstrained frameworks. In this case, we have $n_m > 0$, $n_s = 0$, and $r = n_e < 3\,n_n$. The framework is statically overdetermined. This means that the equilibrium equation (1.3) has a unique solution when $\boldsymbol{f} \in \mathrm{col}(\boldsymbol{A})$, that is, when loads do null work on the mechanisms, otherwise there are no solutions. The framework is also kinematically underdetermined, in that Eq. (1.6) always have a n_m-parameter family of solutions. The equilibrium operator $\boldsymbol{A}$ is represented by a rectangular 'tall' matrix of full rank.

Type 3: overconstrained frameworks. In this case, we have $n_m = 0$, $n_s > 0$, and $r = 3\,n_n < n_e$. The framework is statically underdetermined. This means that the equilibrium equation (1.3) always has a n_s-parameter family of solutions. The framework is also kinematically overdetermined, in that Eq. (1.6) has a unique solution when $\boldsymbol{e} \in \mathrm{col}(\boldsymbol{A}^T)$, that is, when the work done by any selfstress state on the elongations is null, otherwise there are no solutions. The equilibrium operator $\boldsymbol{A}$ is represented by a rectangular 'short' matrix of full rank.

Type 4: simultaneously under- and overconstrained frameworks. In this case, we have $n_m > 0$, $n_s > 0$, and $r < \min\{3\,n_n, n_e\}$. The equilibrium matrix does not have full rank and the statements of the previous cases, Type 2 and Type 3, hold simultaneously. The framework is both statically and kinematically over- and underdetermined: equation (1.3) has a n_s-parameter family of solutions if loads do null work on the mechanisms, otherwise there are no solutions; similarly, equation (1.6) has a n_m-parameter family of solutions if selfstress states do null work on the elongations, otherwise there are no solutions.

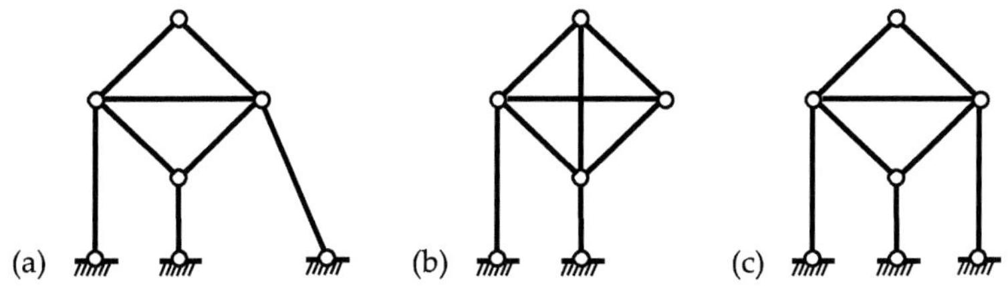

Fig. 1.4 Three systems with the same number of nodes and edges. The Type-1 system **a** can be transformed into a Type-4 system by moving one bar in a different position as in **b**, or by changing the position of its nodes to reach a non-generic configuration

With regard to Type-4 systems, it can be observed that two different situations can be realized, one in which the system is partly underconstrained and partly overconstrained, the other one in which the system has a special geometry. With the terminology introduced in Sect. 1.1.3, a system of Type 1, 2, or 3 is in generic configuration, and it can become a Type-4 system maintaining the same node-edge incidence relation if the configuration is changed to a non-generic one. Figure 1.4 shows three systems with the same number of nodes and edges. The system in Fig. 1.4b is obtained from the Type-1 system in Fig. 1.4a by moving one edge in a different position, which makes it a Type-4 system partially under- and partly overconstrained with $n_s = n_m = 1$. The Type-4 system in Fig. 1.4c is obtained from system in Fig. 1.4a by changing the position of a node so that the three bottom bars are aligned, making the configuration non-generic; again $n_s = n_m = 1$.

Bar frameworks with a non-null stress state can be converted into tensegrity frameworks by replacing bars in tension with cables. Tensegrity frameworks can be found in configurations that belong to any of the four types listed. The typical case is that of Type-3 or Type-4 systems in which cable elements are in tension because of the presence of a certain selfstress state. For Type-1 or Type-2 systems, the stress state is induced by a permanent load. This is the case, for example, of the SpiderBot (Capua et al., 2009) or the Spidercam system, which are underconstrained cable-driven robots controlled by winches in which the cables are held in tension by the gravity load.

1.3.4 Singular-Value Decomposition of the Equilibrium Matrix

Selfstress states and mechanisms can be readily determined by the singular-value decomposition of the equilibrium operator (cf. Pellegrino, 1993), which can be written as follows:

$$\boldsymbol{A} = \boldsymbol{U}\boldsymbol{V}\boldsymbol{W}^T. \tag{1.13}$$

In (1.13), $\boldsymbol{U}$ is given by a square orthogonal matrix of dimension $3\,n_n$. The column vectors of such matrix, $\boldsymbol{u}_i, i = 1, \ldots, 3\,n_n$, are the *left singular vectors* of $\boldsymbol{A}$. The first

r left singular vectors constitute an orthonormal basis of col$(\boldsymbol{A})$ and are collected in $\boldsymbol{U}_r$; the remaining n_m left singular vectors constitute an orthonormal basis of null$(\boldsymbol{A}^T)$, the subspace of mechanisms, and are collected in $\boldsymbol{U}_m$:

$$[\boldsymbol{U}] = [\boldsymbol{u}_1 \ldots \boldsymbol{u}_r \mid \boldsymbol{u}_{r+1} \ldots \boldsymbol{u}_{r+n_m}] = [\boldsymbol{U}_r \mid \boldsymbol{U}_m]\,. \tag{1.14}$$

The operator $\boldsymbol{W}$ is given by a square orthogonal matrix of dimension n_e. The column vectors of such matrix, $\boldsymbol{w}_i, i = 1, \ldots, n_e$, are the right singular vectors of $\boldsymbol{A}$. The first r right singular vectors constitute an orthonormal basis of col$(\boldsymbol{A}^T)$ and are collected in $\boldsymbol{W}_r$; the remaining n_s right singular vectors constitute an orthonormal basis of null$(\boldsymbol{A})$, the subspace of selfstress states, and are collected in $\boldsymbol{W}_s$:

$$[\boldsymbol{W}] = [\boldsymbol{w}_1 \ldots \boldsymbol{w}_r \mid \boldsymbol{w}_{r+1} \ldots \boldsymbol{w}_{r+n_s}] = [\boldsymbol{W}_r \mid \boldsymbol{W}_s]\,. \tag{1.15}$$

The operator $\boldsymbol{V}$ is given by a diagonal matrix having only the first r entries on the diagonal, $v_i, i = 1, \ldots, r$, different from zero. These are the singular values of $\boldsymbol{A}$, which are strictly positive and listed in decreasing order. We observe that the equilibrium operator can be expressed also as

$$\boldsymbol{A} = \boldsymbol{U}_r\, \boldsymbol{V}_r\, \boldsymbol{W}_r^T\,,$$

with $\boldsymbol{V}_r = \text{diag}(v_1, \ldots, v_r)$. The first r left and right singular vectors are in a one-to-one correspondence. In fact, we can write:

$$\boldsymbol{A}\boldsymbol{w}_i = \begin{cases} v_i\,\boldsymbol{u}_i \text{ if } 1 \leq i \leq r \quad (\boldsymbol{w}_i \in \text{col}(\boldsymbol{A}^T)) \\ \quad \boldsymbol{0} \quad \text{if } r < i \leq n_e \ (\boldsymbol{w}_i \in \text{null}(\boldsymbol{A})) \end{cases}; \tag{1.16}$$

$$\boldsymbol{A}^T\boldsymbol{u}_i = \begin{cases} v_i\,\boldsymbol{w}_i \text{ if } 1 \leq i \leq r \qquad (\boldsymbol{u}_i \in \text{col}(\boldsymbol{A})) \\ \quad \boldsymbol{0} \quad \text{if } r < i \leq 3\,n_n \ (\boldsymbol{u}_i \in \text{null}(\boldsymbol{A}^T)) \end{cases}. \tag{1.17}$$

We recall that the singular-value decomposition can be obtained by computing eigenvalues and eigenvectors of the operators $\boldsymbol{A}\boldsymbol{A}^T$ and $\boldsymbol{A}^T\boldsymbol{A}$. It can be seen that these operators have the same non-null eigenvalues, which are also all positive. The square roots of such eigenvalues are the singular values of $\boldsymbol{A}$, while the left and right singular vectors of $\boldsymbol{A}$ are the eigenvalues of $\boldsymbol{A}\boldsymbol{A}^T$ and $\boldsymbol{A}^T\boldsymbol{A}$, respectively.

1.3.5 *The Equilibrium Problem for Linearly Elastic Bar Frameworks*

We here review the equilibrium problem for bar frameworks in linear elasticity. In addition to the equilibrium and kinematic compatibility equations already introduced, we consider a constitutive equation for each edge of the form $t_{ij} = k_{ij}\, e_{ij}$, with k_{ij}

the stiffness constant of edge ij, and write all the constitutive equation in compact form as

$$\boldsymbol{t} = \boldsymbol{D}\boldsymbol{e}\,, \tag{1.18}$$

with $\boldsymbol{D} = \mathrm{diag}(\ldots, k_{ij}, \ldots)$. For simplicity, we disregard inelastic elongations caused by distortions and changes in temperature. We consider a bar framework of Type 1 (isostatic) or Type 3 (overconstrained) initially load-free and unstressed. An elastic state is given by the collection of nodal displacements, elements' elongations, and axial forces, $\mathcal{S} = \{\boldsymbol{u}, \boldsymbol{e}, \boldsymbol{t}\}$ satisfying the kinematic compatibility equation and the constitutive equation. Given the nodal positions, the free and constrained nodes, the incidence relation between the nodes and the edges, the stiffness constants of the edges, and the external nodal loads $\boldsymbol{f}$, the equilibrium problem consists in finding an elastic state satisfying $\boldsymbol{A}\boldsymbol{t} = \boldsymbol{f}$.

In case of a Type 1 system, $\boldsymbol{A}$ is invertible and the solution can be found by solving (1.3), (1.18), $(1.6)_2$ sequentially to determine the elastic state:

$$\boldsymbol{t} = \boldsymbol{A}^{-1}\boldsymbol{f}\,, \qquad \boldsymbol{e} = \boldsymbol{D}^{-1}\boldsymbol{t}\,, \qquad \boldsymbol{u} = (\boldsymbol{A}^T)^{-1}\boldsymbol{e}\,.$$

In case of a Type 3 system, by introducing the stiffness operator $\boldsymbol{K} = \boldsymbol{A}\boldsymbol{D}\boldsymbol{A}^T$, we can express the external load $\boldsymbol{f}$ in terms of $\boldsymbol{u}$,

$$\boldsymbol{f} = \boldsymbol{A}\boldsymbol{t} = \boldsymbol{A}\boldsymbol{D}\boldsymbol{e} = \boldsymbol{A}\boldsymbol{D}\boldsymbol{A}^T\boldsymbol{u} = \boldsymbol{K}\boldsymbol{u}\,.$$

It is easy to see that $\boldsymbol{K}$ is invertible since $\boldsymbol{D}$ is invertible and $\boldsymbol{A}$ is rectangular short of full rank. Therefore, the elastic state is determined as

$$\boldsymbol{u} = \boldsymbol{K}^{-1}\boldsymbol{f}\,, \qquad \boldsymbol{e} = \boldsymbol{A}^T\boldsymbol{u}\,, \qquad \boldsymbol{t} = \boldsymbol{D}\boldsymbol{e}\,.$$

1.4 Stiffness and Stability of Tensegrity Frameworks

A tensegrity framework $\mathcal{T} = \{\mathcal{N}, \mathcal{E}, \mathcal{L}, \mathcal{P}\}$ is given in terms of four sets, other than $\mathcal{N}$ (nodes), $\mathcal{E}$ (edges), and $\mathcal{P}$ (a configuration), it is also specified a labeling $\mathcal{L}$ of the edges, which can be either 'bar', for edges which can carry both tension and compression, or 'cable', for edges which can carry tension only. The axial force of each edge can be expressed as $t_{ij} = k_{ij}(l_{ij} - \bar{l}_{ij})$, $ij \in \mathcal{E}$, with k_{ij} and $\bar{l}_{ij}$ the stiffness constant and the rest length of the edge, respectively. A stress state is *proper* if it is non-negative for cables and *strictly proper* if it is positive for cables, that is, $t_{ij} > 0$, or, $l_{ij} > \bar{l}_{ij}$.

1.4.1 Quasistatic Response and Tangent Stiffness

To keep the analysis simple, we assume that cables remain in tension (strictly proper stress, no cable slackening) and verify a posteriori that this condition is satisfied. We refer the reader to Quirant (2007), Maceri et al. (2014) for analysis methods that take into account the unilateral behavior of cables.

The total potential energy of a tensegrity framework subjected to given constant nodal loads $\boldsymbol{f}$ can be written as follows

$$E_{\text{tot}} = E_{\text{el}} + E_{\text{f}} ,$$

in which E_{el} is the elastic energy, expressed as

$$E_{\text{el}} = \frac{1}{2} \sum_{ij \in \mathcal{E}} k_{ij} \left(l_{ij}(\boldsymbol{p}) - \bar{l}_{ij} \right)^2 , \tag{1.19}$$

and E_{f} is the energy of the loads, expressed as

$$E_{\text{f}} = -\boldsymbol{f} \cdot \boldsymbol{u} .$$

In the former expressions, $\boldsymbol{p}$ is the current configuration and $\boldsymbol{u} = \boldsymbol{p} - \boldsymbol{p}_{\text{init}}$ is the displacement vector from a given initial configuration $\boldsymbol{p}_{\text{init}}$.

An equilibrium configuration $\boldsymbol{p}_{\text{eq}}$ is characterized by the stationary condition

$$\partial_p E_{\text{tot}}(\boldsymbol{p}_{\text{eq}}) = \partial_p E_{\text{el}}(\boldsymbol{p}_{\text{eq}}) - \boldsymbol{f} = \boldsymbol{0} , \tag{1.20}$$

in which we denoted by $\partial_p(\cdot)$ the first derivative with respect to $\boldsymbol{p}$ and considered that $\partial_p (\boldsymbol{f} \cdot \boldsymbol{u}) = \partial_u (\boldsymbol{f} \cdot \boldsymbol{u}) = \boldsymbol{f}$. We will show later that the derivative of the elastic energy is expressed as $\partial_p E_{\text{el}} = \boldsymbol{A}\boldsymbol{t}$, with $\boldsymbol{A}$ the equilibrium operator and $\boldsymbol{t}$ the vector of axial forces. An equilibrium configuration is stable if the potential energy has a minimum at $\boldsymbol{p}_{\text{eq}}$, that is, the second derivative, or Hessian, $\partial_p^2 E_{\text{tot}} = \partial_p^2 E_{\text{el}}$ is positive definite at $\boldsymbol{p}_{\text{eq}}$.

We assume that, in a quasistatic process, for any loading $\boldsymbol{f}$, we can solve (1.20) to find the corresponding equilibrium configuration $\boldsymbol{p}_{\text{eq}}$, so that we can determine the map

$$\boldsymbol{p} \mapsto \boldsymbol{f}(\boldsymbol{p}) = \partial_p E_{\text{el}} .$$

By considering a certain equilibrium configuration, say $\boldsymbol{p}_0$, and by taking the Taylor expansion of $\boldsymbol{f}(\boldsymbol{p})$ about $\boldsymbol{p}_0$,

$$\boldsymbol{f}(\boldsymbol{p}) = \boldsymbol{f}(\boldsymbol{p}_0) + \partial_p \boldsymbol{f}(\boldsymbol{p}_0)\,(\boldsymbol{p} - \boldsymbol{p}_0) + o(|\boldsymbol{p} - \boldsymbol{p}_0|) ,$$

we can express the incremental problem for a small change in configuration induced by a small load increment,

$$\Delta \boldsymbol{f} \cong \boldsymbol{K}_T(\boldsymbol{p}_0)\,\Delta \boldsymbol{p}\,, \tag{1.21}$$

with $\Delta \boldsymbol{f} = \boldsymbol{f}(\boldsymbol{p}) - \boldsymbol{f}(\boldsymbol{p}_0)$, $\Delta \boldsymbol{p} = \boldsymbol{p} - \boldsymbol{p}_0$, and

$$\boldsymbol{K}_T(\boldsymbol{p}_0) := \partial_{\boldsymbol{p}} \boldsymbol{f}(\boldsymbol{p}_0) = \partial^2_{\boldsymbol{p}} E_{\mathrm{el}}(\boldsymbol{p}_0)$$

the *tangent stiffness* operator at $\boldsymbol{p}_0$. Under the assumption that $\boldsymbol{p}_0$ is a stable equilibrium configuration, the change in configuration can be calculated as $\Delta \boldsymbol{p} = \boldsymbol{K}_T^{-1}(\boldsymbol{p}_0)\,\Delta \boldsymbol{f}$.

1.4.2 Tangent Stiffness Calculation

Here we derive explicit expressions of the second derivative of the elastic energy, that is, the tangent stiffness operator, which we will use to define the geometric stiffness operator and to give the notions of prestress stability and superstability.

We start by considering the first and second derivatives of the length of an edge. From the expression

$$l_{ij} = |\boldsymbol{p}_i - \boldsymbol{p}_j| = \sqrt{(x_i - x_j)^2 + (y_i - y_j)^2 + (z_i - z_j)^2}\,,$$

since $\partial_{x_i} l_{ij} = \frac{1}{l_{ij}}(x_i - x_j)$ and $\partial_{x_j} l_{ij} = -\frac{1}{l_{ij}}(x_i - x_j)$, it can be seen that,

$$\partial_{\boldsymbol{p}} l_{ij} = \frac{1}{l_{ij}} \boldsymbol{C}_{ij} \boldsymbol{p}\,, \tag{1.22}$$

with $\boldsymbol{C}_{ij}$ the connectivity operator of edge ij, represented by a boolean $3n_n$-by-$3n_n$ block matrix whose non-null 3-by-3 blocks are in positions ii, ij, ji, and jj,

$$[\boldsymbol{C}_{ij}] = \begin{bmatrix} & & & \\ & \mathbf{1}_3 & -\mathbf{1}_3 & \\ & & & \\ & -\mathbf{1}_3 & \mathbf{1}_3 & \\ & & & \end{bmatrix},$$

where $\mathbf{1}_3$ denotes the three-dimensional identity operator.

On comparing (1.22) with (1.5) and (1.8), it can also be seen that the first derivative of the length of edge ij is equal to the corresponding column vector $\boldsymbol{a}_{ij} \in \mathcal{V}^{3n_n}$ of the equilibrium operator $\boldsymbol{A}$, which has non-null entries only in positions i and j:

$$[\partial_{p} l_{ij}] = \frac{1}{l_{ij}} \begin{bmatrix} \boldsymbol{p}_i - \boldsymbol{p}_j \\ \\ \boldsymbol{p}_j - \boldsymbol{p}_i \end{bmatrix} = \begin{bmatrix} \boldsymbol{n}_{ji} \\ \\ \boldsymbol{n}_{ij} \end{bmatrix} = [\boldsymbol{a}_{ij}]\,. \tag{1.23}$$

The second derivative of the edge length can then be calculated as

$$\partial^2_{\boldsymbol{p}} l_{ij} = -\frac{1}{l^2_{ij}} (\boldsymbol{C}_{ij}\boldsymbol{p})(\partial_{\boldsymbol{p}} l_{ij})^T + \frac{1}{l_{ij}} \boldsymbol{C}_{ij} = \frac{1}{l_{ij}} (-\boldsymbol{a}_{ij}\, \boldsymbol{a}_{ij}^T + \boldsymbol{C}_{ij})\,,$$

where used the product rule $\partial_{\boldsymbol{p}}(\alpha \boldsymbol{v}) = \boldsymbol{v}(\partial_{\boldsymbol{p}}\alpha)^T + \alpha\, \partial_{\boldsymbol{p}} \boldsymbol{v}$, with α a scalar and $\boldsymbol{v}$ a vector.

We can now calculate the first derivative of the elastic energy. From (1.19), we have,

$$\partial_{\boldsymbol{p}} E_{\text{el}} = \sum_{ij \in \mathcal{E}} k_{ij}(l_{ij}(\boldsymbol{p}) - \bar{l}_{ij})\, \partial_{\boldsymbol{p}} l_{ij}(\boldsymbol{p}) = \sum_{ij \in \mathcal{E}} \frac{t_{ij}}{l_{ij}} \boldsymbol{C}_{ij}\boldsymbol{p} = \sum_{ij \in \mathcal{E}} t_{ij}\, \boldsymbol{a}_{ij} = \boldsymbol{A}\boldsymbol{t}\,. \tag{1.24}$$

The second derivative of the energy, or tangent stiffness operator, can be calculated as follows,

$$\begin{aligned} \boldsymbol{K}_T = \partial^2_{\boldsymbol{p}} E_{\text{el}} &= \sum_{ij \in \mathcal{E}} \left(k_{ij} (\partial_{\boldsymbol{p}} l_{ij})(\partial_{\boldsymbol{p}} l_{ij})^T + t_{ij}\, \partial^2_{\boldsymbol{p}} l_{ij} \right) \\ &= \sum_{ij \in \mathcal{E}} \left(k_{ij}\, \boldsymbol{a}_{ij}\, \boldsymbol{a}_{ij}^T + \frac{t_{ij}}{l_{ij}} (-\boldsymbol{a}_{ij}\, \boldsymbol{a}_{ij}^T + \boldsymbol{C}_{ij}) \right). \end{aligned}$$

This expression can be rewritten in the form

$$\boldsymbol{K}_T = \boldsymbol{A}\boldsymbol{D}\boldsymbol{A}^T + \boldsymbol{S} - \boldsymbol{A}\boldsymbol{Q}\boldsymbol{A}^T \tag{1.25}$$

that is obtained by recalling the definition of force densities $q_{ij} = t_{ij}/l_{ij}$ and by introducing the diagonal matrices $\boldsymbol{D} = \text{diag}(\ldots, k_{ij}, \ldots)$, $\boldsymbol{Q} = \text{diag}(\ldots, q_{ij}, \ldots)$, while

$$\boldsymbol{S} = \sum_{ij \in \mathcal{E}} q_{ij}\, \boldsymbol{C}_{ij} \tag{1.26}$$

is the so-called *stress operator*, which has a central role in form-finding problems. The expression (1.25) shows that the tangent stiffness operator is the sum of the material stiffness $\boldsymbol{K}_M = \boldsymbol{A}\boldsymbol{D}\boldsymbol{A}^T$, equal to the linear stiffness operator introduced in Sect. 1.3.5, and the geometric stiffness $\boldsymbol{K}_G = \boldsymbol{S} - \boldsymbol{A}\boldsymbol{Q}\boldsymbol{A}^T$.

It is worth writing explicitly these two operators for the typical element that connects nodes $P, Q \in \mathcal{E}$. The configuration vector $\boldsymbol{p}^{(e)}$ of such element contains the position vectors of these nodes

$$[\boldsymbol{p}^{(e)}] = \begin{bmatrix} \boldsymbol{p}_P \\ \boldsymbol{p}_Q \end{bmatrix} .$$

The tangent stiffness operator of this element $\boldsymbol{K}_T^{(e)}$ relates the load increment

$$[\Delta \boldsymbol{f}^{(e)}] = \begin{bmatrix} \Delta \boldsymbol{f}_P \\ \Delta \boldsymbol{f}_Q \end{bmatrix}$$

to the configuration change $\Delta \boldsymbol{p}^{(e)}$,

$$\Delta \boldsymbol{f}^{(e)} = \boldsymbol{K}_T^{(e)} \, \Delta \boldsymbol{p}^{(e)} ,$$

with $\boldsymbol{K}_T^{(e)}$ expressed as the block matrix

$$[\boldsymbol{K}_T^{(e)}] = \begin{bmatrix} \boldsymbol{K} & -\boldsymbol{K} \\ -\boldsymbol{K} & \boldsymbol{K} \end{bmatrix} , \tag{1.27}$$

where

$$\boldsymbol{K} = k_{PQ} \, \boldsymbol{n}_{PQ} \, \boldsymbol{n}_{PQ}^T + q_{PQ} \, (\mathbf{1}_3 - \boldsymbol{n}_{PQ} \, \boldsymbol{n}_{PQ}^T) , \tag{1.28}$$

k_{PQ}, q_{PQ} being the stiffness constant and force density of the element. The expression (1.27) can be used for the standard finite-element assemblage of the tangent stiffness of the entire framework.

We note that, when $\boldsymbol{K}_T^{(e)}$ is applied to $\Delta \boldsymbol{p}^{(e)}$, $\boldsymbol{K}$ is applied to the displacement vector of node P or node Q. Since the operator $\boldsymbol{n}_{PQ} \, \boldsymbol{n}_{PQ}^T$ projects vectors along the direction of $\boldsymbol{n}_{PQ}$, and $(\mathbf{1} - \boldsymbol{n}_{PQ} \, \boldsymbol{n}_{PQ}^T)$ projects vectors on the plane orthogonal to $\boldsymbol{n}_{PQ}$, it follows that the first term in (1.28) (material stiffness) comes into play when the edge changes length, while the second term in (1.28) (geometric stiffness) comes into play when the edge changes direction. We also notice that by substituting the constitutive relation in the expression of the force density

$$q_{PQ} = \frac{k_{PQ}}{l_{PQ}} (l_{PQ} - \bar{l}_{PQ}) = k_{PQ} \left(1 - \frac{\bar{l}_{PQ}}{l_{PQ}} \right) ,$$

we have

$$\boldsymbol{K} = k_{PQ} \frac{\bar{l}_{PQ}}{l_{PQ}} \, \boldsymbol{n}_{PQ} \, \boldsymbol{n}_{PQ}^T + q_{PQ} \, \mathbf{1}_3 \cong k_{PQ} \, \boldsymbol{n}_{PQ} \, \boldsymbol{n}_{PQ}^T + q_{PQ} \, \mathbf{1}_3 ,$$

where the last approximation applies to the case of very stiff edges, $\bar{l}_{PQ} \cong l_{PQ}$. In this case, the corresponding tangent stiffness for the entire framework can be approximated as

$$\boldsymbol{K}_T \cong \boldsymbol{A} \boldsymbol{D} \boldsymbol{A}^T + \boldsymbol{S} .$$

We conclude this subsection by observing that, in view of (1.24) and (1.26), the equilibrium equations can also be expressed in terms of the stress operator $\boldsymbol{S}$,

$$\boldsymbol{f} = \boldsymbol{A}\,\boldsymbol{t} = \widehat{\boldsymbol{A}}\,\boldsymbol{q} = \boldsymbol{S}\,\boldsymbol{p}\,, \tag{1.29}$$

from which it is evident their bilinear dependence on the force densities, $\boldsymbol{q}$, and nodal coordinates, $\boldsymbol{p}$.

1.4.3 Prestress Stability and Superstability

There are two specialized notions of stability for tensegrity systems: prestress stability and superstability. The former applies to prestressed systems possessing mechanisms (Type 4, see Sect. 1.3.3), the latter can be applied to any prestressed system (Type 3 or Type 4).

Let us consider the incremental problem (1.21) at a selfstressed equilibrium configuration $\boldsymbol{p}_0$ of a system with mechanisms. We define the *geometric load* vector $\boldsymbol{g}_m$ as the vector that contains the nodal loads associated with a configuration change along a mechanism $\boldsymbol{u}_m$; by taking $\Delta\boldsymbol{p} = \boldsymbol{u}_m$ in (1.21), we have

$$\boldsymbol{g}_m := \boldsymbol{K}_T(\boldsymbol{p}_0)\boldsymbol{u}_m = (\boldsymbol{A}(\boldsymbol{D} - \boldsymbol{Q})\boldsymbol{A}^T + \boldsymbol{S})\boldsymbol{u}_m = \boldsymbol{S}\,\boldsymbol{u}_m\,, \tag{1.30}$$

in which the last equality follows by taking into account that, for a mechanism, $\boldsymbol{A}^T\boldsymbol{u}_m = \boldsymbol{0}$. The configuration $\boldsymbol{p}_0$ is prestress stable if the work done by the geometric load on every mechanism is positive,

$$\boldsymbol{g}_m \cdot \boldsymbol{u}_m > 0 \quad \forall \boldsymbol{u}_m \in \text{null}(\boldsymbol{A}^T)\,, \tag{1.31}$$

that is,

$$\boldsymbol{K}_T\,\boldsymbol{u}_m \cdot \boldsymbol{u}_m \;=\; \boldsymbol{S}\,\boldsymbol{u}_m \cdot \boldsymbol{u}_m > 0 \quad \forall \boldsymbol{u}_m \in \text{null}(\boldsymbol{A}^T)\,. \tag{1.32}$$

By recalling the singular value decomposition of the equilibrium matrix (1.13), any mechanism of the framework can be expressed as a linear combination of the last n_m left singular vectors, collected in the operator $\boldsymbol{U}_m$,

$$\boldsymbol{u}_m = \sum_{i=1}^{n_m} \alpha_i\,\boldsymbol{u}_{r+i} = \boldsymbol{U}_m\boldsymbol{\alpha}\,,$$

with $\boldsymbol{\alpha}$ the vector containing the weights of the combination, α_i, $i = 1, \ldots, n_m$. Thus, the prestress stability condition (1.32) can be rewritten as

$$\boldsymbol{S}\,\boldsymbol{u}_m \cdot \boldsymbol{u}_m = \boldsymbol{S}\boldsymbol{U}_m\boldsymbol{\alpha} \cdot \boldsymbol{U}_m\boldsymbol{\alpha} = \boldsymbol{U}_m^T\boldsymbol{S}\boldsymbol{U}_m\boldsymbol{\alpha} \cdot \boldsymbol{\alpha} = \boldsymbol{S}_m\boldsymbol{\alpha} \cdot \boldsymbol{\alpha} > 0 \quad \forall \boldsymbol{\alpha}\,,$$

which states that the operator $\boldsymbol{S}_m = \boldsymbol{U}_m^T \boldsymbol{S} \boldsymbol{U}_m$, of dimensions $n_m \times n_m$, obtained by reducing the stress operator to the subspace of mechanisms, should be positive definite. We note that the definition of prestress stability given above applies to the case $n_s = 1$ (only one independent selfstress state); if $n_s > 1$, then any chosen linear combination of selfstress states can be tested.

Prestress stability ensures that every mechanism (for which the material stiffness is null) is associated with a positive geometric stiffness. For this reason, we say that the selfstress state stiffens the mechanisms of the system. However, prestress stability does not imply stability. In fact, even if the prestress stability condition is satisfied, it can happen that the geometric stiffness is negative for certain configuration changes that are not mechanisms. A prestressed system in equilibrium, with or without mechanisms, is superstable if and only if: (1) it is prestress stable, in case there are mechanisms, and (2) the geometric stiffness operator $\boldsymbol{K}_G$ is positive semidefinite. Superstability implies that the tangent stiffness operator $\boldsymbol{K}_T$ is positive definite independently of the material properties k_{ij}, $ij \in \mathcal{E}$, and of the scaling of the prestress level (excluding a sign change).

We remark that the definition of superstability given above is different from the definition usually given in the literature (Connelly, 2002) (cf. also Zhang & Ohsaki, 2007; Micheletti et al., 2019), which states that a system is superstable if the following conditions are satisfied: (i) the stress matrix is positive definite; (ii) the stress matrix has corank equal to 12; (iii) the system has no affine motions. The definition given here is equivalent to that in Connelly (2002) in the case of positive rest lengths, but it also makes sense for the case of negative rest lengths, which may occur when elastic cable reel devices are considered (Micheletti, 2013).

1.5 Nonlinear Quasistatic Response

In this section, we describe some basic examples to better understand the nonlinear behavior of tensegrity systems under quasistatic loads.

1.5.1 Geometric Response

Consider the two-dimensional Type-2 system shown in Fig. 1.5 (left) composed of just one vertical rigid bar of length l hinged at the top end and subjected to a vertical downward force f_0 (preload) at the bottom end. Obviously, the axial force in the bar is $t_0 = f_0$. The bar can rotate about the hinge and every configuration is described by the rotation angle θ counted counterclockwise from the vertical position. We want to determine the equilibrium configurations when the bottom end point is subjected to a horizontal load increment Δf. We observe that there would be no equilibrium solution for a horizontal loading when the bar is vertical according to the linear theory. However, an equilibrium configuration can be reached by a change of configuration.

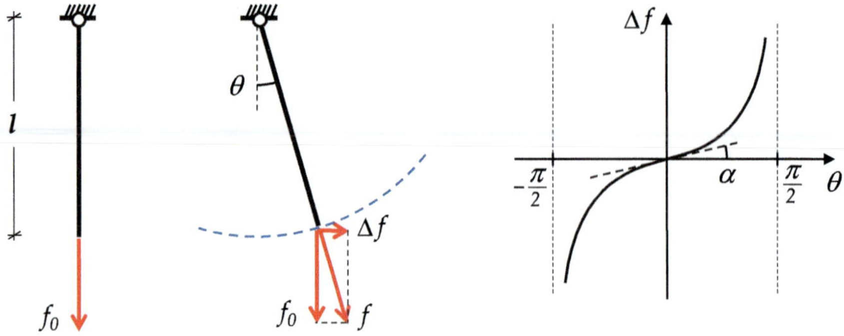

Fig. 1.5 A preloaded system composed by just one bar subjected to a load increment (see description in the main text)

In this case, the system is in equilibrium when the bar and the resultant load f are collinear (Fig. 1.5, center), so that the resultant moment with respect to the hinged point is null. The axial force in the bar at equilibrium is then $t = f$. The exact relationship between the load increment and θ is $\Delta f = f_0 \tan\theta$ and it is plotted in Fig. 1.5 (right). We notice that the slope of the curve at the origin is $\tan\alpha = f_0$ and that $\Delta f \to \pm\infty$ when $\theta \to \pm\pi/2$.

Consider now small-magnitude load increments such that θ is sufficiently small to write $\tan\theta \cong \theta$. The horizontal displacement of the bottom end of the bar is then $u = l \sin\theta \cong l\theta$. The relationship between the small horizontal load increment and the small horizontal displacement is

$$\Delta f = f_0\,\theta = f_0\,\frac{u}{l} = \frac{t_0}{l}\,u = q_0\,u\,,$$

with $q_0 = t_0/l$ the force coefficient of the bar. This calculation shows that the force coefficient q_0 measures the initial stiffness of the system with respect to the load increment and that such stiffness is only related to the geometric effect of the change in configuration.

1.5.2 First-Order Infinitesimal Mechanisms

We now look at the two-dimensional Type-4 system in Fig. 1.6a composed of two horizontal equal linearly elastic bars of length l_0, which has one selfstress state and one internal mechanism. The bars carry the axial force $t_0 = k\,(l_0 - \bar{l})$, with $\bar{l}$ their rest length. The internal mechanism corresponds to a vertical displacement of the center node. We want to study the response of the system to a vertical load f applied to the center node. We remark that there would be no equilibrium solution for such a

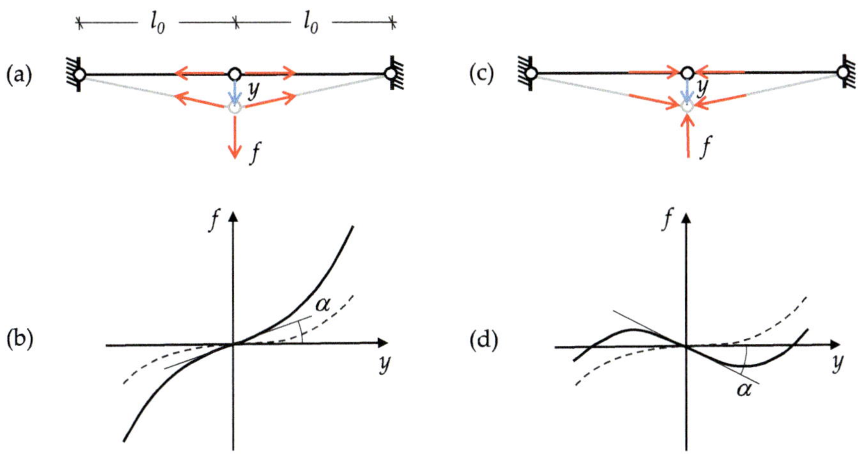

Fig. 1.6 Different load-versus-displacement responses of a two-bar system (see description in the main text)

loading when the bars are horizontal according to the linear theory, but an equilibrium solution can be found in a different configuration.

By considering the symmetry of this problem, we write the total potential energy of the system as a function of the vertical coordinate y of the center node counted from the initial configuration,

$$E(y) = E_{el} + E_f = k\,(l(y) - \bar{l})^2 - f\,y\,,$$

where the current length $l(y) = \sqrt{l_0^2 + y^2}$. The equilibrium condition is expressed as

$$0 = E'(y) = 2\,k\,(l(y) - \bar{l})\,l'(y) - f\,. \tag{1.33}$$

Since $t(y) = k\,(l(y) - \bar{l})$ and $l'(y) = y/l(y) = \sin\theta$, with θ the rotation angle of each bar, we obtain the equilibrium relationship $f = 2\,t\,\sin\theta$. For each given f we can determine the equilibrium value of y to define the map $y \mapsto f(y) = E'_{el}$, which associates with every y the vertical force that keeps the system in equilibrium.

The second derivative of the energy, which also measures the tangent stiffness, reads

$$E''(y) = f'(y) = 2\,k\,(l'^2 + (l - \bar{l})\,l'') = 2\,k\left(1 - \frac{\bar{l}}{l}\left(1 - \frac{y^2}{l^2}\right)\right). \tag{1.34}$$

It can be seen that $E''(y) > 0$ if $l_0 > \bar{l}$, which means that the equilibrium configuration is always stable if the two bars are initially in tension. The initial stiffness has the expression

$$f'(0) = 2k\left(1 - \frac{\bar{l}}{l_0}\right) = 2\frac{t_0}{l_0} = 2q_0\,,$$

from which we see that each bar contributes to it with its force coefficient q_0.

Since $l(y)$ is an even function, $l(y) = l_0 + O(y^2)$: elongations do not undergo first-order changes with respect to y and $\Delta l = l - l_0 = O(y^2)$. Systems with this property have been referred to as *first-order infinitesimal mechanisms* by Pellegrino and Calladine (1986). In the same way, $t(y) = t_0 + O(y^2)$, so that the axial forces do not change at the first order. Moreover, since the equilibrium load f is expressed as a derivative of an even function, we have $f(y) = 2\,q_0\,y + O(y^3)$, so that the force-versus-displacement relationship can be approximated by a cubic with an inflection point at the origin (Fig. 1.6b). The slope at the origin, $\tan\alpha$, is directly proportional to the selfstressing force t_0. As in the previous example, for small loads the equilibrium is realized geometrically by a change of configuration. If $t_0 = 0$, the initial stiffness is null, the linear approximation does not apply, and even a small load would cause a large displacement (dashed curve in Fig. 1.6b).

Taking into account the case $l_0 < \bar{l}$, corresponding to $t_0 < 0$ (Fig. 1.6c), condition (1.33) determines three load-free equilibrium configurations, two for $y = \pm(\bar{l}^{\,2} - l_0^2)^{-1/2}$, corresponding to $l(y) = \bar{l}$, and one for $y = 0$ with collinear bars in compression. We see from (1.34) that the latter is unstable and that the former are stable. The response of the system to the vertical load is now described by the curve shown in Fig. 1.6d that is typical of a bistable system.

On looking at the prestress stability condition (1.31), for the case of Fig. 1.6a the geometric load $\boldsymbol{g}_m$ is a vertical downward force of magnitude $2q_0\,y$ and does positive work for the displacement of the center node, so that the collinear configuration is prestress stable. For the case of Fig. 1.6c the geometric load in this case is a vertical upward force that does negative work on the displacement of the center node and the prestress stability condition is not satisfied.

The properties of this two-bar system are analogous to the properties of more complex elastic tensegrity structures that are first-order infinitesimal mechanisms. Oppenheim and Williams (2000) studied the quasistatic response of tensegrity prisms describing their highly nonlinear response. Moreover, Fraternali et al. (2014a) demonstrated that tensegrity prisms can be designed to have extreme stiffening or softening behavior in response to applied loads, a fact that was exploited to develop tensegrity metamaterials supporting the propagation of different types of solitary waves (Fraternali et al., 2014b). In addition, Vangelatos et al. (2020) and Intrigila et al. (2022) showed that it is possible to design bistable tensegrity-like structures based on prestress unstable tensegrity systems with $n_s = 1$ and $n_m = 1$, while Bauer et al. (2021) presented a tensegrity-like metamaterial with increased resistance to failure through delocalized deformation.

Finally, it is worth stressing that if external loads cause a cable to go slack, then the load-versus-displacement curve would show a drop in stiffness marked by a discontinuity on the first derivative.

1.5.3 Bistable Regimes

In the previous example, we observed how by changing the sign of selfstress it is possible to switch from a monostable system to a bistable system. This switch can also occur by a change in geometry, without changes in the sign of the axial forces, as shown by Micheletti (2013) and Micheletti et al. (2023). In those studies, the configuration of the system passes from prestress-stable to prestress-unstable, and two other stable equilibrium configurations emerge. The snapping motion between the two stable configurations takes place along the direction of an the internal mechanism. In addition, it was shown that it is also possible to have a second bistable regime (Micheletti, 2013; Micheletti et al., 2023). In fact, a prestress stable but not superstable configuration can become unstable above a certain selfstress threshold, with a possible monostable-to-bistable switching.

Figure 1.7 summarizes the findings obtained in Micheletti et al. (2023). The eight-node system in Fig. 1.7a can exhibit stable configurations with symmetry point group D_{2h} (left configuration), when it is monostable, and with symmetry point group D_2

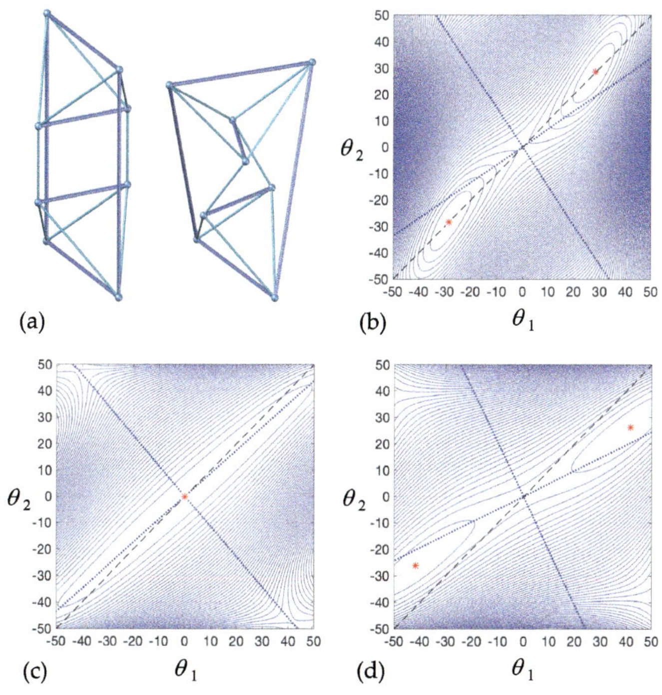

Fig. 1.7 **a** Different stable configurations of a eight-node system (Micheletti et al., 2023). **b–d** Elastic energy contour plots: **b** prestress unstable case; **c** prestress stable case with selfstress level below the threshold; **d** prestress stable case with selfstress level above the threshold. The dashed line indicates the direction of the relevant internal mechanism, and the dotted lines indicate the directions corresponding to the eigenvalues of the tangent stiffness operator

(right configuration), when it is bistable. Figure 1.7b–d show the contour plot of the elastic energy in terms of the configuration parameters θ_1, θ_2 that measure the relative rotation angles about the vertical axis between, respectively, the outer and inner pairs of horizontal bars. The dashed line indicates the direction of the relevant internal mechanism, while the dotted lines indicate the directions of the eigenvalues of the tangent stiffness operator. When the geometry of the right configuration in Fig. 1.7a is such that the system is prestress unstable, the energy landscape takes the form shown in Fig. 1.7b: the system is bistable with energy minima located along the direction of the internal mechanism. When the geometry of the right configuration in Fig. 1.7a is such that the system is prestress stable and the selfstress level is below a certain threshold, the corresponding energy landscape is the one shown in Fig. 1.7c: the system is monostable. If the selfstress level exceeds that threshold, the prestress stable system becomes unstable with the energy landscape shown in Fig. 1.7d: the system becomes bistable with the energy minima not located along the direction of the mechanism.

1.6 Dynamic Response

In this section, we review the motion equation for tensegrity systems in both its nonlinear and linearized versions. We consider two types of damping force, one linear in the nodal velocities and the other linear in the elongation rates of the edges, which we refer to as natural damping. We recall that, as shown by Oppenheim & Williams (2001), if the system constitutes a first-order infinitesimal mechanism and only natural damping is present, then the energy decay of such a system is slower than that of a linearly damped system.

1.6.1 Motion Equation

The energy dissipation rate given by a viscous damping force associated with nodal velocities can be expressed as

$$\mathcal{D}^{(p)} = \sum_{i\in\mathcal{N}} d_i^{(p)} |\dot{\boldsymbol{p}}_i|^2 = \boldsymbol{D}^{(p)} \dot{\boldsymbol{p}} \cdot \dot{\boldsymbol{p}} \,,$$

with $\boldsymbol{D}^{(p)}$ a $3n_n$-by-$3n_n$ block-diagonal matrix whose i-th block on the diagonal is $d_i^{(p)}\mathbf{1}$, that is, $\boldsymbol{D}^{(p)} = \text{blockdiag}(\ldots, d_i^{(p)}\mathbf{1}, \ldots)$, with $d_i^{(p)}$ the damping constant for the i-th node. The energy dissipation rate associated to the elements' rate of elongation (natural damping) is

$$\mathcal{D}^{(l)} = \sum_{ij\in\mathcal{E}} d_{ij}^{(l)} \dot{l}_{ij}^2 = \sum_{ij\in\mathcal{E}} d_{ij}^{(l)} (\partial_p l_{ij} \cdot \dot{\boldsymbol{p}})^2 = \sum_{ij\in\mathcal{E}} d_{ij}^{(l)} (\boldsymbol{a}_{ij} \cdot \dot{\boldsymbol{p}})^2 = \boldsymbol{A}\boldsymbol{D}^{(l)}\boldsymbol{A}^T \dot{\boldsymbol{p}} \cdot \dot{\boldsymbol{p}} \,,$$

where $\boldsymbol{D}^{(l)} = \text{diag}(\ldots, d_{ij}^{(l)}, \ldots)$, with $d_{ij}^{(l)}$ the damping constant of edge ij. Then, the total energy dissipation rate is

$$\mathcal{D} = \mathcal{D}^{(p)} + \mathcal{D}^{(l)} \,.$$

The energy balance can be written as $\dot{E} + \mathcal{D} = 0$, in which $\dot{E}$ is the total energy rate of change, with $E = E_{el}(\boldsymbol{p}) + E_f(\boldsymbol{p}) + E_{kin}(\dot{\boldsymbol{p}})$. The terms E_{el} and E_f are the elastic energy and the potential energy of the loads previously introduced, while E_{kin} is the kinetic energy,

$$E_{kin} = \frac{1}{2} \boldsymbol{M} \dot{\boldsymbol{p}} \cdot \dot{\boldsymbol{p}} \,,$$

expressed in terms of a constant, symmetric mass operator $\boldsymbol{M}$. The energy rate of change can be computed as

$$\dot{E} = \partial_{\boldsymbol{p}} E_{el} \cdot \dot{\boldsymbol{p}} + \partial_{\boldsymbol{p}} E_f \cdot \dot{\boldsymbol{p}} + \partial_{\dot{\boldsymbol{p}}} E_{kin} \cdot \ddot{\boldsymbol{p}} \,,$$

so that the energy balance can be rewritten as

$$\dot{E} + \mathcal{D} = \boldsymbol{A}\boldsymbol{t} \cdot \dot{\boldsymbol{p}} - \boldsymbol{f} \cdot \dot{\boldsymbol{p}} + \boldsymbol{M}\ddot{\boldsymbol{p}} \cdot \dot{\boldsymbol{p}} + (\boldsymbol{A}\boldsymbol{D}^{(l)}\boldsymbol{A}^T + \boldsymbol{D}^{(p)})\dot{\boldsymbol{p}} \cdot \dot{\boldsymbol{p}} = 0 \,.$$

By introducing the operator $\boldsymbol{C}_d = \boldsymbol{A}\boldsymbol{D}^{(l)}\boldsymbol{A}^T + \boldsymbol{D}^{(p)}$, after collecting $\dot{\boldsymbol{p}}$, we get

$$(\boldsymbol{A}\boldsymbol{t} - \boldsymbol{f} + \boldsymbol{M}\ddot{\boldsymbol{p}} + \boldsymbol{C}_d\dot{\boldsymbol{p}}) \cdot \dot{\boldsymbol{p}} = 0 \,,$$

from which we get the motion equation

$$\boldsymbol{M}\ddot{\boldsymbol{p}} + \boldsymbol{C}_d(\boldsymbol{p})\dot{\boldsymbol{p}} + \boldsymbol{A}(\boldsymbol{p})\boldsymbol{t}(\boldsymbol{p}) = \boldsymbol{f} \,. \tag{1.35}$$

We recall here two typical expressions for the mass operator of an edge, say $PQ \in \mathcal{E}$, which can be used for the finite-element assemblage of the mass operator of the entire system. With the notation introduced in Sect. 1.4.2, the kinetic energy of the edge is $E_{kin}^{(e)} = \frac{1}{2}\boldsymbol{M}^{(e)}\dot{\boldsymbol{p}}^{(e)} \cdot \boldsymbol{p}^{(e)}$. On denoting by m_{PQ} the total mass of the edge and considering that such mass is lumped in equal fractions at the nodes P and Q, we have

$$[\boldsymbol{M}^{(e)}] = m_{PQ} \begin{bmatrix} \frac{1}{2}\boldsymbol{1}_3 & \boldsymbol{0}_3 \\ \boldsymbol{0}_3 & \frac{1}{2}\boldsymbol{1}_3 \end{bmatrix} ,$$

with $\mathbf{0}_3$ a null 3-by-3 matrix. In case the mass is uniformly distributed along the edge, we have

$$[\boldsymbol{M}^{(e)}] = m_{PQ} \begin{bmatrix} \frac{1}{3}\mathbf{1}_3 & \frac{1}{6}\mathbf{1}_3 \\ \frac{1}{6}\mathbf{1}_3 & \frac{1}{3}\mathbf{1}_3 \end{bmatrix} .$$

1.6.2 Linearized Dynamics

Let us consider a stable equilibrium configuration $\boldsymbol{p}_0$ under the load $\boldsymbol{f}_0$,

$$\boldsymbol{A}(\boldsymbol{p}_0)\boldsymbol{t}_0 = \boldsymbol{f}_0$$

with $\boldsymbol{t}_0 = \boldsymbol{t}(\boldsymbol{p}_0)$. We have

$$\begin{aligned} \boldsymbol{A}\boldsymbol{t} &= \partial_{\boldsymbol{p}} E_{el}(\boldsymbol{p}) = \partial_{\boldsymbol{p}} E_{el}(\boldsymbol{p}_0) + \partial^2_{\boldsymbol{p}} E(\boldsymbol{p}_0)(\boldsymbol{p} - \boldsymbol{p}_0) + O(|\boldsymbol{p} - \boldsymbol{p}_0|^2) \\ &= \boldsymbol{A}(\boldsymbol{p}_0)\boldsymbol{t}_0 + \boldsymbol{K}_T(\boldsymbol{p}_0)\boldsymbol{u} + O(|\boldsymbol{u}|^2) = \boldsymbol{f}_0 + \boldsymbol{K}_T(\boldsymbol{p}_0)\boldsymbol{u} + O(|\boldsymbol{u}|^2) , \end{aligned}$$

with $\boldsymbol{u} = \boldsymbol{p} - \boldsymbol{p}_0$. Then, by replacing the expression above in (1.35), and by retaining only the first-order terms, we obtain the linearized equation of motion for small oscillations about the configuration $\boldsymbol{p}_0$,

$$\boldsymbol{M}\ddot{\boldsymbol{u}} + \boldsymbol{C}_d(\boldsymbol{p}_0)\dot{\boldsymbol{u}} + \boldsymbol{K}_T(\boldsymbol{p}_0)\boldsymbol{u} = \Delta\boldsymbol{f} ,$$

where we considered that $\dot{\boldsymbol{u}} = \dot{\boldsymbol{p}}$ and $\Delta\boldsymbol{f} = \boldsymbol{f} - \boldsymbol{f}_0$.

The natural frequencies and vibration modes of the undamped system at p_0 can be found by solving the generalized eigenvalue problem

$$(\boldsymbol{K}_T - \omega^2 \boldsymbol{M})\, Œ = \mathbf{0} .$$

It can be shown that the natural frequency of vibration modes occurring along first-order infinitesimal mechanism scales with the square root of the selftstress level.

1.7 Form-Finding Problems

In this section, after giving some generalities about form-finding problems for tensegrity systems, we describe in detail three useful and relatively simple approaches: an analytical method for symmetric systems, the so-called cocoon method, and a marching algorithm.

A tensegrity system can be specified in terms of three mathematical ingredients (see Fig. 1.8):

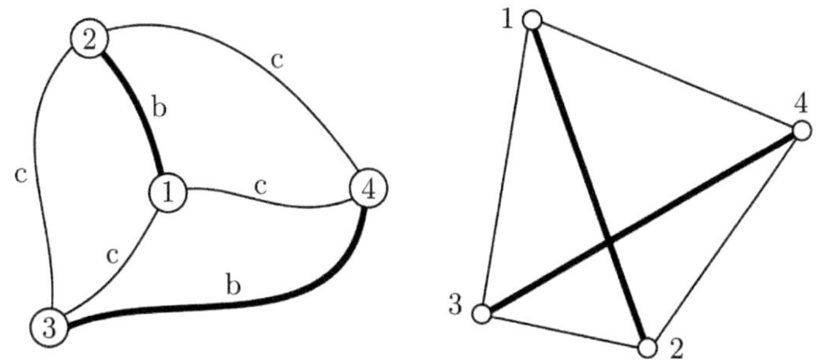

Fig. 1.8 The underlying graph and the labeling of a tensegrity system (left) and one of its stable embeddings with proper selfstress (right)

(i) the underlying graph $\mathcal{G} = (\mathcal{N}, \mathcal{E})$ (node-edge incidence relationship);
(ii) a labeling $\mathcal{L}$ of the edges (cable/bar labels);
(iii) an embedding $\mathcal{P}$ of the graph (nodal positions).
Moreover, a tensegrity system should take at least one configuration in which
(iv) the stress is proper (non-negative in cables);
(v) the configuration is stable (the total potential energy attains a local minimum).
Different form-finding problems can be stated according to what is assumed to be known and what is sought. A typical form-finding problem is that of determining the embedding(s) (iii) for a given graph (i) and labeling (ii) such that conditions (iv) and (v) are satisfied. Conversely, a different form-finding problem consists in finding the possible graphs (i) and labelings (ii) for a given assignment of the nodal positions (iii). In practical design problems, one could also consider a partial assignment of (i), (ii), and/or (iii), and the objective of form finding is to determine the missing information.

The difficulties of form-finding problems are primarily those related to the solution of the equilibrium equations, which are bilinear in the nodal coordinates and force coefficients. Many form-finding algorithms establish iterative calculations based on the singular value decomposition of the operator $\boldsymbol{A}$ and the eigenvalue decomposition of the stress operator $\boldsymbol{S}$ (see, e.g., Zhang & Ohsaki, 2006; Gomez Estrada et al., 2006; Tran & Lee, 2010). A concise review of form-finding methods can be found in Micheletti and Podio-Guidugli (2022).

1.7.1 Analytical Form Finding of a Symmetric System

To exemplify this method, we consider the case of a diamond truncated octahedron, a classic tensegrity system with 24 nodes, 12 bars, and 48 cables that form six squares and eight triangles. Figure 1.9a depicts an instance of this structure.

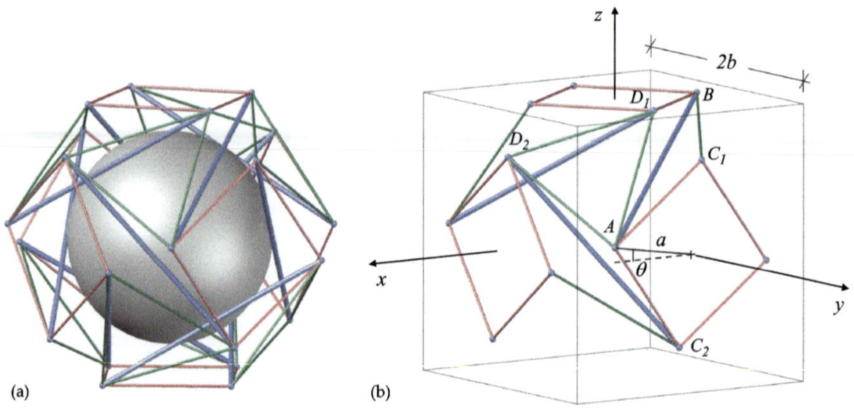

Fig. 1.9 **a** A diamond truncated octahedron (a sphere is drawn inside the structures for ease of visualization). **b** Subsets of nodes and elements of the same structure with the relevant geometric parameters

Table 1.1 Nodal coordinates of the points in Fig. 1.9 (right)

Node	A	B	C_1	C_2	D_1	D_2
x	$a\cos\theta$	$-a\cos\theta$	$-a\sin\theta$	$a\sin\theta$	$a\sin\theta$	b
y	b	$a\sin\theta$	b	b	$a\cos\theta$	$a\sin\theta$
z	$a\sin\theta$	b	$a\cos\theta$	$-a\cos\theta$	b	$a\cos\theta$

The extended Maxwell's rule (1.12) applied to this system gives $n_m - n_s = 6$. We assume that there is at least one selfstress state and look for a non-null solution of the equilibrium equation with null external loads, $\boldsymbol{At} = \boldsymbol{0}$. We consider only configurations with the symmetries of the cube and, in addition, with six four-fold and eight three-fold cyclic-symmetry axes. Under this assumption, the equilibrium equation of one node is equivalent to the equilibrium equation of any other node. Figure 1.9b shows subset of elements with the geometric parameters that identify any configuration of the system. All nodes lies on the faces of a cube of side length $2b$. The distance between a node and the center of the corresponding face is a. A Cartesian reference frame centered and aligned with the cube axes is chosen. The angle θ measures the twist of the squares of cables clockwise; with $\theta = 0$ when all nodes lie on the Cartesian planes.

We consider the equilibrium equation of the typical node A shown in Fig. 1.9b, written in terms of the force densities. The coordinates of the relevant nodes are reported in Table 1.1. The equilibrium equation of node A reads as follows

$$q_{AB}(\boldsymbol{p}_A - \boldsymbol{p}_B) + q_{AC_1}(\boldsymbol{p}_A - \boldsymbol{p}_{C_1}) + q_{AC_2}(\boldsymbol{p}_A - \boldsymbol{p}_{C_2}) + q_{AD_1}(\boldsymbol{p}_A - \boldsymbol{p}_{D_1}) + q_{AD_2}(\boldsymbol{p}_A - \boldsymbol{p}_{D_2}) = \boldsymbol{0}.$$

Under the symmetry assumption $q_{AC_1} = q_{AC_2}, q_{AD_1} = q_{AD_2}$, and we can rewrite the equilibrium condition in matrix form as

$$\begin{bmatrix} 2ac & 2ac & -b+2ac-as \\ b-as & 0 & 2b-ac-as \\ -b+as & 2as & -b-ac+2as \end{bmatrix} \begin{bmatrix} q_{AB} \\ q_{AC_1} \\ q_{AD_1} \end{bmatrix} = [\mathbf{0}] \,,$$

where we replaced $c = \cos\theta$ and $s = \sin\theta$.

To have non-null solutions, the determinant of the coefficient matrix must vanish. Dividing the first and third column of the matrix by b, the second column by $2a$, and substituting $\gamma := a/b$, this condition becomes:

$$\det \begin{bmatrix} 2\gamma c & c & -1+2\gamma c-\gamma s \\ 1-\gamma s & 0 & 2-\gamma c-\gamma s \\ -1+\gamma s & s & -1-\gamma c+2\gamma s \end{bmatrix} = 0 \,.$$

A straightforward calculation permit to obtain the following analytical condition:

$$\gamma^2 \sin^2\theta\,(\cos\theta + \sin\theta) + 2\gamma\cos\theta\,(\cos\theta - \sin\theta) - \cos\theta - \sin\theta = 0 \,.$$

Meaningful form-finding solutions, with a proper selfstress and prestress-stable, are found for $0 < \gamma < 1$ and $-\pi/4 < \theta < \pi/4$, with $\gamma \to 0$ when $\theta \to -\pi/4$ and $\gamma \to 1$ when $\theta \to \pi/4$. A particular solution is $\theta = 0$, $\gamma = 1/2$.

1.7.2 Virtual Cocoon Method

This method is based on the balloon analogy and on the approach followed by Sakantamis and Popovic Larsen (2004). The idea consists in inserting a certain number of bars into a cable-net envelope and arranging them to push the envelope outward in different directions to obtain a spatial structure. By elongating the bars or shrinking the cable net, a stable configuration is found. Figure 1.10 shows a two-dimensional example. Sakantamis and Popovic Larsen (2004) realized several physical models in which the envelope is initially composed of a membrane, and once a desired shape is obtained, cables are positioned along the directions of higher tension and the membrane is removed. In Micheletti and Cadoni (2011) and Cadoni and Micheletti (2012), the approach of Sakantamis and Popovic Larsen (2004) is reproduced by numerical simulations of elastic tensegrity balloons. Here, we present some examples of this procedure. Once the type of cable net envelope and the disposition of a certain number of internal bars are chosen, together with a stress-free and load-free starting configuration, the rest length of the bars is iteratively increased and the elastic energy of the system (1.19) is minimized by standard numerical routines.

Figure 1.11a shows the result of the form finding for a cable net in which three pairs of bars are inserted along three mutually orthogonal directions. In Fig. 1.11b

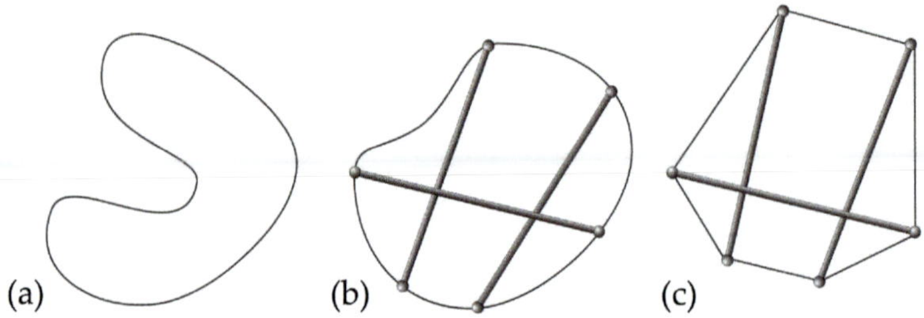

Fig. 1.10 The cocoon method applied to a two-dimensional system. Three bars are inserted in the envelope **a** to obtain the system **b**. Stable selfstressed configurations as in **c** can be obtained by shortening the cables and/or by lengthening the bars

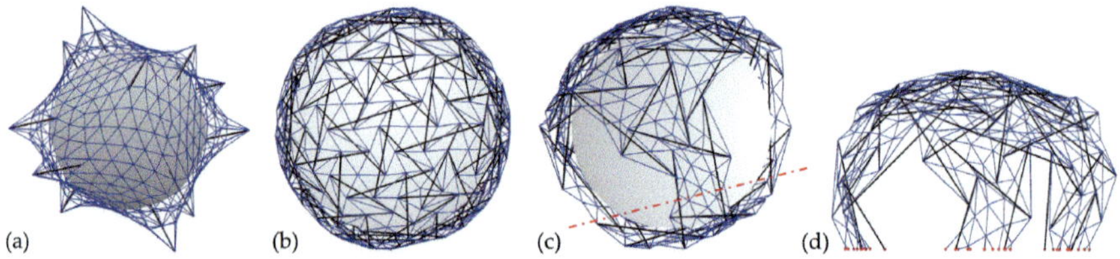

Fig. 1.11 Examples of application of the virtual cocoon method. **a** Form finding of a cable net in which three pairs of bars are inserted along three mutually orthogonal directions. **b** A tensegrity balloon with a triangulated cable net and 120 bars. **c** a tensegrity balloon in which the cable net has large holes. **d** A tensegrity canopy obtained by sectioning the structure in **c** and pin-connecting the sectioned elements to the ground. A sphere is drawn inside the structures **a**–**c** for ease of visualization

a tensegrity balloon with a triangulated cable net and 120 bars is shown. The bar arrangement mimics that of a truncated icosahedral tensegrity structure. Figure 1.11c shows a tensegrity balloon in which the cable net has large holes. Figure 1.11d illustrates an application of the method to the design of a tensegrity canopy obtained by sectioning the structure in Fig. 1.11c and pin-connecting the sectioned elements to the ground.

Figure 1.12 shows the results of the method for structures based on cable-bar polygons. Figure 1.12a, b shows two views of a seed structure obtained by expanding a cable-bar polygon in the out-of-plane direction using an additional bar and a few extra cables. Figure 1.12c, d show two views of a large tensegrity balloon obtained in a similar way from several cable-bar polygons laying in parallel planes. Figure 1.12e, f show a computer model and a physical model of a tensegrity canopy realized by sectioning the previous structure.

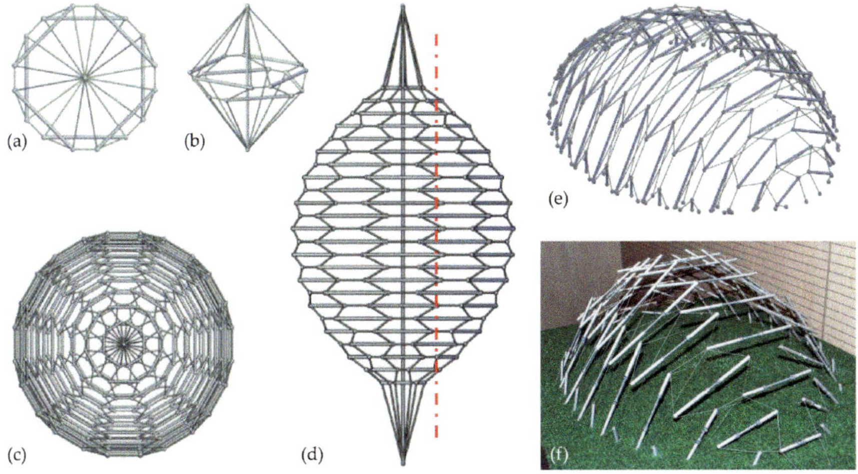

Fig. 1.12 Tensegrity balloons and a tensegrity canopy based on cable-bar polygons. **a**, **b** Plan view and side view of a simple unit. **c**, **d** Plan view and side view of a tensegrity balloon obtained form several cable-bar polygons arranged on parallel planes. **e** A tensegrity canopy obtained by sectioning the structure in **c, d**. **f** Photo of a desktop model built by Davide Cadoni

1.7.3 Shape-Change by Controlling Edge Lengths

We take the cue from Micheletti & Williams (2007) to describe this method. Consider the set $\mathcal{R}_r$ of configurations $\boldsymbol{p}$ for which the rank of the equilibrium operator $\widehat{\boldsymbol{A}}(\boldsymbol{p})$ is r, with r not maximal (Type-4 configurations). The collection of all these configurations forms a smooth manifold and we can write

$$\mathcal{R}_r = \{\boldsymbol{p} : \text{rank}(\widehat{\boldsymbol{A}}(\boldsymbol{p})) = r < \min\{3n_n - 6, n_e\}\}\,.$$

We assume that we have a starting configuration $\boldsymbol{p}_0$ at which the prestress stability condition (1.32) holds. By the continuity of the nullspaces, there exists a neighborhood where the stability condition is satisfied. We consider how to construct paths on the manifold that remain within this neighborhood. Following Micheletti & Williams (2007), it can be shown that at every point $\boldsymbol{p}$ on $\mathcal{R}_r$, the tangent space is spanned by all the vectors that are normal to the geometric loads $\boldsymbol{g}_m$ for all the choices of self-stresses $\boldsymbol{t}_s$ and mechanisms $\boldsymbol{u}_m$. Moreover, the projections of the vectors $\widehat{\boldsymbol{a}}_{ij}(\boldsymbol{p})$ (the column vectors of $\widehat{\boldsymbol{A}}$) on the tangent space span this space. Since the vectors $\widehat{\boldsymbol{a}}_{ij}(\boldsymbol{p})$ are linearly dependent by assumption, we can choose an independent subset of these vectors, and since $\widehat{\boldsymbol{a}}_{ij}(\boldsymbol{p})$ also represents the gradient of $\frac{1}{2}l_{ij}^2(\boldsymbol{p})$ (cf. (1.23)),

$$\partial_{\boldsymbol{p}}(\tfrac{1}{2}l_{ij}^2) = l_{ij}\partial_{\boldsymbol{p}}l_{ij} = l_{ij}\,\boldsymbol{a}_{ij} = \widehat{\boldsymbol{a}}_{ij}\,,$$

the corresponding edge lengths will form a local coordinate system on the manifold.

Consider now a motion of the structure $\boldsymbol{p}(t)$ as a time-parameterized family of configurations, starting at the stable configuration $\boldsymbol{p}_0 = \boldsymbol{p}(0)$. We use the above characterization to set up a differential system of equations for $\boldsymbol{p}(t)$, whose solution is a motion composed only of stable configurations. Firstly, if a motion is on the manifold, the velocity vector $\dot{\boldsymbol{p}}(t)$ belongs to the tangent space, that is, it is orthogonal to all the geometric loads:

$$\boldsymbol{g}_m \cdot \dot{\boldsymbol{p}} = 0\,. \tag{1.36}$$

Then, to prescribe the lengthenings of the edges, we can use the compatibility equations (1.7) written as follows

$$\widehat{\boldsymbol{a}}_{ij}(\boldsymbol{p}) \cdot \dot{\boldsymbol{p}} = \widehat{e}_{ij}\,, \tag{1.37}$$

with $\widehat{\boldsymbol{a}}_{ij}$ the column of $\widehat{\boldsymbol{A}}$ corresponding to $ij \in \mathcal{E}$.

The differential system obtained from equations (1.36) and (1.37) can be solved efficiently using the Runge-Kutta integration algorithm. To test the accuracy of results, the singular values of the equilibrium operator can be monitored. The number of the last singular values that are numerically close to zero corresponds to the corank of the equilibirum operator, $(\min\{3n_n - 6, n_e\} - r)$. The ratio ρ_{sv} between the first singular value close to zero and the previous one can be taken as a measure of how close the configuration is to the rank deficiency manifold.

A two-dimensional example. Consider the system in Fig. 1.13, solid line, which has one mechanism and one selfstress state. We want to decrease the length of the edge AB while keeping BD and AC with constant length and remaining on the manifold; the length of the edge BE is changed accordingly. The nodes A and B will then follow the circular trajectories shown, obtaining the dashed line configuration in Fig. 1.13. The path on the manifold $\boldsymbol{p}(t)$ is the solution of the following differential system:

$$\begin{cases} \widehat{\boldsymbol{a}}_{AB} \cdot \dot{\boldsymbol{p}} = \widehat{e}_{AB} < 0 \\ \widehat{\boldsymbol{a}}_{AC} \cdot \dot{\boldsymbol{p}} = 0 \\ \widehat{\boldsymbol{a}}_{BD} \cdot \dot{\boldsymbol{p}} = 0 \\ \boldsymbol{g}_m \cdot \dot{\boldsymbol{p}} = 0 \end{cases},$$

which can also be written as

$$\begin{cases} (\boldsymbol{p}_A - \boldsymbol{p}_B) \cdot (\dot{\boldsymbol{p}}_A - \dot{\boldsymbol{p}}_B) = \widehat{e}_{AB} < 0 \\ (\boldsymbol{p}_A - \boldsymbol{p}_C) \cdot (\dot{\boldsymbol{p}}_A - \dot{\boldsymbol{p}}_C) = 0 \\ (\boldsymbol{p}_B - \boldsymbol{p}_D) \cdot (\dot{\boldsymbol{p}}_B - \dot{\boldsymbol{p}}_D) = 0 \\ \boldsymbol{g}_m \cdot \dot{\boldsymbol{p}} = 0 \end{cases}.$$

Tensegrity prism. Consider the symmetric tensegrity prism with triangular base shown in Fig. 1.14 (right). We recall that the upper base is rotated with respect to the lower base by an angle of $\pi/6$. This system has one internal mechanisms and one selfstress state. As in the previous example, we move on the manifold by changing the length of two edges by increasing the length of BD and finding the length of

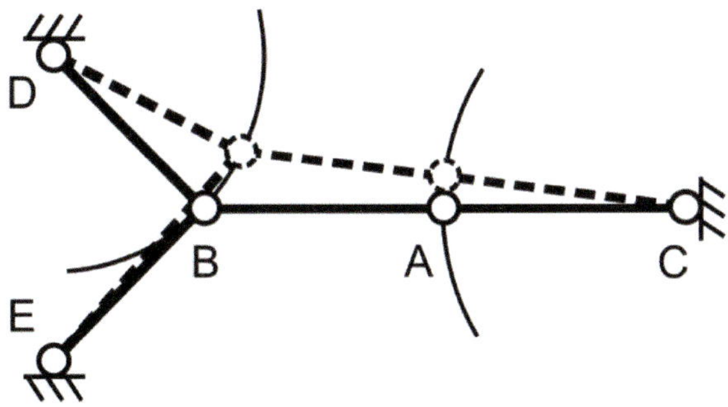

Fig. 1.13 Shape-change of a simple two-dimensional system of Type 4. The edges AB and BE change length, while the configuration remains on the rank-deficiency manifold

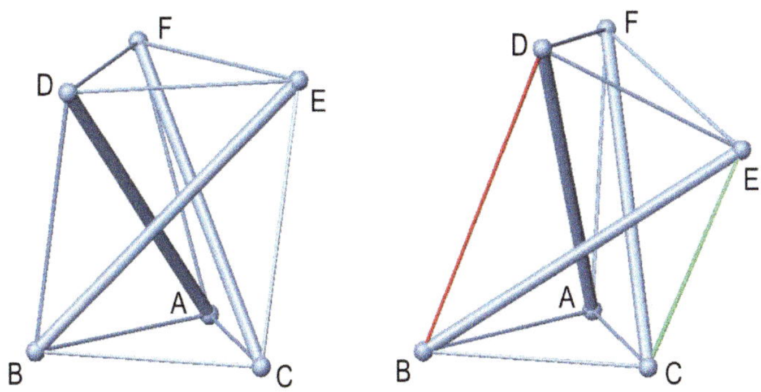

Fig. 1.14 Shape-change of a triangular tensegrity prism

CE to remain on the manifold. The length of all the other edges stays the same. The system of differential equations is set as follows:

$$\begin{cases} \widehat{a}_{BD} \cdot \dot{p} = \widehat{e}_{BD} > 0 \\ \widehat{a}_{ij} \cdot \dot{p} = 0, \quad ij \neq BD, CE \\ g_m \cdot \dot{p} = 0 \end{cases} .$$

We numerically solved this system of equations and determined the singular values of the equilibrium operator using the Matlab^{TM} built-in routines *ode45* and *svd*, respectively. At the starting configuration, the circumscribed circle to the two equal horizontal triangular bases has a diameter of 20 units; the height of the module is 20 units. The lateral cables have the same length, $l_{AF} = l_{BD} = l_{CE} \cong 20.659$ units. In this analytically determined configuration, the ratio ρ_{sv} is of the order of 10^{-16}. In the final configuration, shown in Fig. (1.14) (right), we have $l_{BD} \cong 25.036$ units, $l_{CE} \cong 17.232$ units, and $\rho_{sv} \sim 10^{-12}$.

Double prism A double prism can be obtained by superposing two tensegrity prisms, as shown in Fig. 1.15 (left): the lower node of the upper prism are connected to the mid

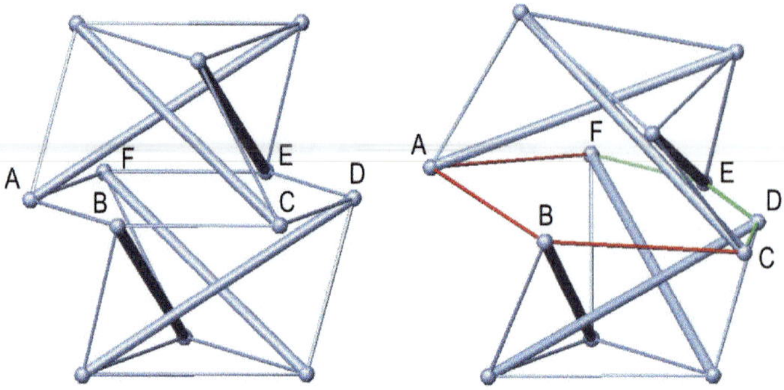

Fig. 1.15 Shape-change of a double tensegrity prism

points of the upper cables of the lower prism. In a stable equilibrium configuration, these cables lie in a plane forming a hexagon, and the system has one selfstress state and seven independent mechanisms. Consequently, there are seven independent geometric load vectors g_h, $h = 1, \dots, 7$. Starting at a symmetric configuration, three contiguous edges of the hexagon are lengthened at the same rate (AF, AB, and BC); while remaining on the manifold, the other three change length at the same rate (CD, DE, and EF). The length of the other edges remains unchanged. The differential system of equations reads as follows:

$$\begin{cases} \widehat{\boldsymbol{a}}_{AF} \cdot \dot{\boldsymbol{p}} = \widehat{e}_{AF} > 0 \\ (\widehat{\boldsymbol{a}}_{AF} - \widehat{\boldsymbol{a}}_{AB}) \cdot \dot{\boldsymbol{p}} = 0 \\ (\widehat{\boldsymbol{a}}_{AB} - \widehat{\boldsymbol{a}}_{BC}) \cdot \dot{\boldsymbol{p}} = 0 \\ (\widehat{\boldsymbol{a}}_{CD} - \widehat{\boldsymbol{a}}_{DE}) \cdot \dot{\boldsymbol{p}} = 0 \\ (\widehat{\boldsymbol{a}}_{DE} - \widehat{\boldsymbol{a}}_{EF}) \cdot \dot{\boldsymbol{p}} = 0 \\ \boldsymbol{g}_h \cdot \dot{\boldsymbol{p}} = 0\,, \quad h = 1, \dots, 7 \end{cases} .$$

In this system, the 2nd and 3rd equations follow from the conditions $\widehat{e}_{AF} = \widehat{e}_{AB}$, $\widehat{e}_{AB} = \widehat{e}_{BC}$. Similarly, the 4th and 5th equations impose the same rate of elongation to edges CD, DE, and EF but the actual rate is determined by the solution of the system. For this process, we have $\rho_{sv} \sim 10^{-15}$ at the starting configuration and $\rho_{sv} \sim 10^{-10}$ in the final one shown in Fig. 1.15 (right).

1.8 Case Study of a Tensegrity Ring for Deployable Mesh Reflectors

Mesh reflectors are large aperture telecommunication antennas that are deployed in orbit. They are made up of a deployable structure that supports a paraboloid-shaped cable net, which in turn supports a fine mesh layer that constitutes the reflective

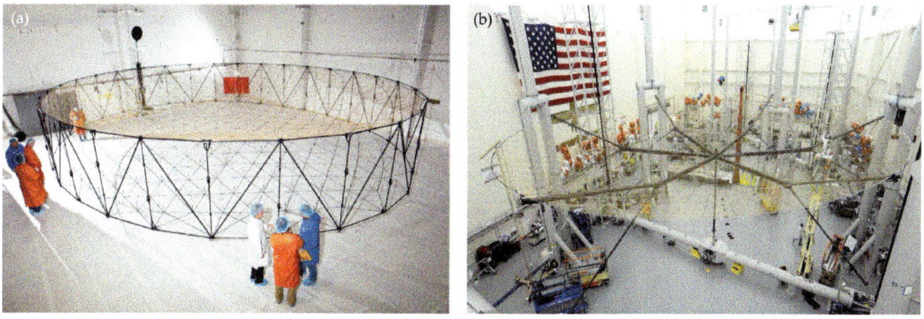

Fig. 1.16 **a** The 12-m Astromesh reflector. **b** The 18-m Terrestar reflector. Adapted with permission from Ganga et al. (2016)

surface of the antenna. There are two deployment architectures that have been successfully developed in the past years: the expandable peripheral ring and the hinged radial rib architectures (Fig. 1.16). In the former, the peripheral ring applies tension to two counterposed paraboloidal triangulated cable nets, which give shape and pretension to the reflective mesh layer. In hinged radial rib reflectors, rigid radial elements in compression apply tension to a paraboloidal cable net by means of cable ties, also in this case giving shape and pretension to the reflective mesh layer.

In Ganga et al. (2016), a new tensegrity structure was proposed for the peripheral ring architecture, realized as a two-stage Snelson tower (Micheletti, 2003). This structure is obtained by superposing two tensegrity prisms, as in the double prism shown in Fig. 1.15 but with the addition of other cables connecting the bars of one prism to the bars of the other prism, as shown in Fig. 1.17. Following Skelton's terminology (Skelton et al., 2002), the elements of this system are distinguished as follows. The side cables of the original prisms are called *vertical* cables, the cables separating the two prisms are called *saddle* cables, the additional cables are called *diagonal* cables, those at the base of the prisms are called *polygon* cables. It is worth remarking that tensegrity towers are superstable, a feature that guarantees the stability of the system while performing configuration changes.

We recall that a tensegrity prism, or tensegrity frustum, has two polygonal bases with n sides and its geometry in cyclic-symmetric configurations can be identified by four parameters (Fig. 1.18): the distances between the center and one vertex of the bottom and top polygons, a and b, the distance between bases, h, and the relative rotation angle between the bases, or twist, φ, which is equal to zero when the bars are vertical. Given the lengths of all the elements, it is possible to assemble a tensegrity frustum in two different configurations, mirror images of each other, which we refer to as configurations with opposite orientations. It is well-known that the solution of the form-finding problem of cyclic symmetric frustums prescribes only the value of the twist to be

$$\varphi = \frac{\pi}{2} + \frac{\pi}{n} =: \varphi_0 .$$

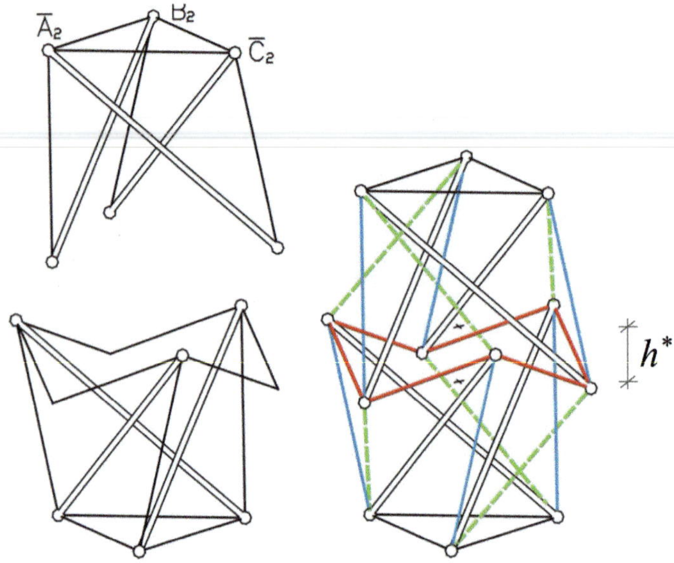

Fig. 1.17 Assembly of a two-stage Snelson tower by superposing two tensegrity prisms, as in the double prism shown in Fig. 1.15, additional cables (dashed lines) connect the bars of one prism to the bars of the other prisms

Fig. 1.18 Geometric parameters of a tensegrity frustum

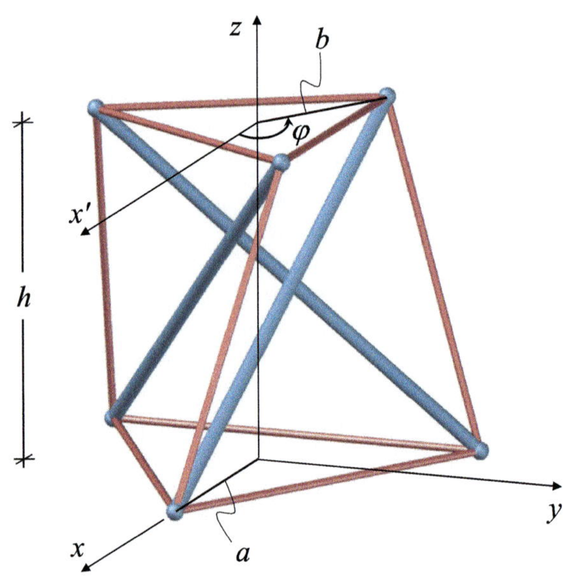

To perform the ring design, we consider a two-stage Snelson tower obtained from two frustums with equal geometry and opposite orientation. Any configuration of such a structure is determined by the four parameters of a frustum plus the overlap distance, h^*, which measure the interpenetration between frustums (Fig. 1.17, right).

We make use of the analytical solution to the form-finding problem for this structure given in Micheletti (2003) in the form

$$\delta^2(\gamma \sin\varphi_0 + \sin\theta) + \delta(\gamma - \gamma \sin\varphi_0 + 2\sin\varphi_0 \sin\theta - 2\sin\theta) - 2\sin\varphi_0 \sin\theta + 2\sin\theta = 0\,, \tag{1.38}$$

where $\theta = \varphi - \varphi_0$, $\delta = h^*/h$, and $\gamma = a/b$. It can be seen that the δ-versus-φ relation for fixed values of n and γ is monotonically decreasing and that ring-shaped structures are obtained for small twists and large overlaps (Ganga et al., 2016).

The design of deployable mesh reflectors must satisfy several technical specifications. The aperture size and focal length determine the dimensions of the paraboloid and, together with the maximum allowed stowed dimensions, constitute the main parameters guiding the design of geometry of the structure and the deploying process. The minimum value of the tension of the reflective mesh layer is a crucial parameter on which the proper functioning of the antenna depends. The required tension of the mesh imposes a certain stress state in the cable net, which in turn determines the stress state of the supporting ring structure. In addition, at the end of the deployment, the structure must have a certain stiffness to ensure proper operating conditions, especially during antenna repositioning, a requirement which imposes the minimum natural frequencies the reflector structure must have. A certain minimal stiffness is also required in the deployment phase.

The above design requirements dictate the geometry of the tensegrity ring and its selfstress state. Here, we focus on the shape-change process and refer the reader to Ganga et al. (2016) for the details of the parametric analysis and design of the ring. The reflector has diameter $D = 12$ m and height $H = 2.6$ m. The latter dimension is sufficient to accommodate two counterposed paraboloids of 1.25 m in height separated by a gap of 0.1 m. The chosen geometric parameters of the two-stage Snelson tower are

$$n = 12\,, \qquad \varphi = 28\,\text{deg}\,, \qquad \gamma = 0.98\,,$$

which, in order to satisfy (1.38), correspond to a value of $\delta \cong 0.8816$. Taking $a = D/2 = 6$ m, we get $b = a/\gamma \cong 6.12$ m and considering that $H = (2 - \delta)h$, we have $h \cong 2.32$ m. The configuration of the tensegrity ring corresponding to these parameters is shown in Fig. 1.19 (left).

The procedure described in the previous section is applied to simulate the folding process. We assign the same elongation rate to all polygon cables to reduce their length to 10% of their deployed value. All vertical cables are imposed to have the same elongation rate while remaining on the rank-deficiency manifold. This tensegrity structure has one selfstress state and a number of mechanisms $n_m = 19$, and this is also the number of independent geometric load vectors, $\boldsymbol{g}_i$, $i = 1, \ldots, 19$, computed according to (1.30). Then, the differential system of equations reads

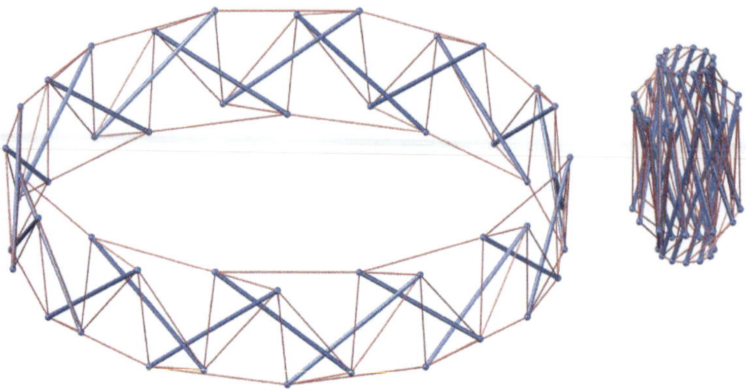

Fig. 1.19 Deployed and folded configuration of the designed tensegrity ring

$$
\begin{cases}
\widehat{a}_{ij} \cdot \dot{p} = \widehat{e} < 0\,, \quad ij \in \mathcal{E}_p \\
(\widehat{a}_{ij} - \widehat{a}_{hk}) \cdot \dot{p} = 0\,, \quad hk \in \mathcal{E}_v\,,\ hk \neq ij \in \mathcal{E}_v \\
\widehat{a}_{ij} \cdot \dot{p} = 0\,, \quad ij \in \mathcal{E}_b \cup \mathcal{E}_s \cup \mathcal{E}_d \\
g_i \cdot \dot{p}\dot{p} = 0\,, \quad i = 1, \ldots, 19
\end{cases}
,
$$

with $\mathcal{E}_p$, $\mathcal{E}_v$, $\mathcal{E}_d$, $\mathcal{E}_s$, $\mathcal{E}_b$ denoting, respectively, the edge subsets of polygon, vertical, diagonal, saddle cables, and bars. Figure 1.19 (right) shows the resulting folded configuration, with a height of about 4.54 m and a diameter of about 2.33 m.

References

Amendola, A., Hernández-Nava, E., Goodall, R., Todd, I., Skelton, R. E., & Fraternali, F. (2015). On the additive manufacturing, post-tensioning and testing of bi-material tensegrity structures. *Composite Structures, 131*, 66–71. https://doi.org/10.1016/j.compstruct.2015.04.038

Amendola, A., Krushynska, A., Daraio, C., Pugno, N. M., & Fraternali, F. (2018). Tuning frequency band gaps of tensegrity metamaterials with local and global prestress. *International Journal of Solids and Structures, 155*, 47–56. https://doi.org/10.1016/j.ijsolstr.2018.07.002

Bauer, J., Kraus, J. A., Crook, C., Rimoli, J. J., & Valdevit, L. (2021). Tensegrity metamaterials: toward failure-resistant engineering systems through delocalized deformation. *Advanced Materials, 33*(10), 2005647. https://doi.org/10.1002/adma.202005647

Bruce, J., Caluwaerts, K., Iscen, A., Sabelhaus, A. P., & SunSpiral, V. (2014). Design and evolution of a modular tensegrity robot platform. In *2014 IEEE international conference on robotics and automation (ICRA)* (pp. 3483–3489). https://doi.org/10.1109/ICRA.2014.6907361

Cadoni, D., & Micheletti, A. (2012). Structural performances of single-layer tensegrity domes. *International Journal of Space Structures, 2–3*, 167–178. https://doi.org/10.1260/0266-3511.27.2-3.167

Calladine, C. R. (1978). Buckminster Fuller's 'tensegrity' structures and Clerk Maxwell's rules for the construction of stiff frames. *International Journal of Solids and Structures, 14*, 161–172. https://doi.org/10.1016/0020-7683(78)90052-5

Capua, A., Shapiro, A., & Shoval. S. (2009). Motion analysis of an underconstrained cable suspended mobile robot. In *Proceedings of the IEEE international conference on robotics and biomimetics. Guilin, China. December 19–23, 2009*.
Connelly, R. (1982). Rigidity and energy. *Inventiones Mathematicae, 66*, 11–33. https://doi.org/10.1007/BF01404753
Connelly, R. (2002). Rigidity theory and applications. In *Chapter tensegrity structures: Why are they stable?* (pp. 47–54). US, Boston, MA: Springer. https://doi.org/10.1007/0-306-47089-6_3
Connelly, R., & Guest, S. D. (2022). *Frameworks, tensegrities, and symmetry*. Cambridge University Press. https://doi.org/10.1017/9780511843297
Connelly, R., & Whiteley, W. (1996). Second-order rigidity and prestress stability for tensegrity frameworks. *SIAM Journal of Discrete Mathematics, 9*, 453–491. https://doi.org/10.1137/S0895480192229236
Davini, C., Micheletti, A., & Podio-Guidugli, P. (2016). On the impulsive dynamics of T3 tensegrity chains. *Meccanica, 51*(11), 2763–2776. https://doi.org/10.1007/s11012-016-0495-y
Feron, J., Boucher, L., Denoël, V., & Latteur, P. (2019). Optimization of footbridges composed of prismatic tensegrity modules. *Journal of Bridge Engineering, 24*(12), 04019112. https://doi.org/10.1061/(ASCE)BE.1943-5592.0001438
Fraternali, F., Senatore, L., & Daraio, C. (2012). Solitary waves on tensegrity lattices. *Journal of the Mechanics and Physics of Solids, 60*(6), 1137–1144. https://doi.org/10.1016/j.jmps.2012.02.007
Fraternali, F., Carpentieri, G., & Amendola, A. (2014). On the mechanical modeling of the extreme softening/stiffening response of axially loaded tensegrity prisms. *Journal of the Mechanics and Physics of Solids, 74*, 136–157. https://doi.org/10.1016/j.jmps.2014.10.010
Fraternali, F., Carpentieri, G., Amendola, A., Skelton, R. E., & Nesterenko, V. F. (2014). Multiscale tunability of solitary wave dynamics in tensegrity metamaterials. *Applied Physics Letters, 105*, 201903. https://doi.org/10.1063/1.4902071
Ganga, P. L., Micheletti, A., Podio-Guidugli, P., Scolamiero, L., Tibert, A. G., & Zolesi, V. (2016). Variational analysis and aerospace engineering: Mathematical challenges for the aerospace of the future. In *Chapter tensegrity rings for deployable space antennas: Concept, design, analysis, and prototype testing* (pp. 269–304). Cham: Springer International Publishing. ISBN 978-3-319-45680-5.
Geiger, D., Stefaniuk, A., & Chen. D. (1986). The design and construction of two cable domes for the korean olympics. In *Proceedings of the IASS symposium, Osaka, Japan, 1986*.
Gilewski, W., & Al Sabouni-Zawadzka, A. (2020). Equivalent mechanical properties of tensegrity truss structures with self-stress included. *European Journal of Mechanics - A/Solids, 83*, 103998. ISSN 0997-7538. https://doi.org/10.1016/j.euromechsol.2020.103998
Gomez Estrada, G., Bungartz, H. J., & Mohrdieck, C. (2006). Numerical form-finding of tensegrity structures. *International Journal of Solids and Structures, 43*(22–23), 6855–6868. https://doi.org/10.1016/j.ijsolstr.2006.02.012
Gómez-Jáuregui, V. (2009). Controversial origins of tensegrity. In Domingo, A., & Lazaro, C (Eds.), *Proceedings of the IASS symposium 2009, Valencia, Spain*. Spain: Universidad Politecnica de Valencia.
Gómez-Jáuregui, V., Carrillo-Rodriguez, A., Manchado, C., & Lastra-Gonzalez, P. (2023). Tensegrity applications to architecture, engineering and robotics: A review. *Applied Sciences, 13*, 8669. https://doi.org/10.3390/app13158669
Hanaor, A. (2012). Debunking "tensegrity" - a personal perspective. *International Journal of Space Structures, 27*, 179–184. https://doi.org/10.1260/0266-3511.27.2-3.179
Ingber, D. E. (1998). The architecture of life. *Scientific American, 278*(1), 48–57. https://doi.org/10.1038/scientificamerican0198-48
Intrigila, C., Micheletti, A., Nodargi, N., Artioli, E., Fraternali, E., & Bisegna, P. (2022). Fabrication and experimental characterization of a bistable tensegrity-like unit for lattice metamaterials. *Submitted*.

Kelly, S. D., Micheletti, A., & Tiero, A. (2019). On the design of superstable prestressed frameworks. *Frontiers in Materials, 6*, 40. https://doi.org/10.3389/fmats.2019.00040

Kim, K., Moon, D., Bin, J. Y., & Agogino, A. M. (2017). Design of a spherical tensegrity robot for dynamic locomotion. In *2017 IEEE/RSJ international conference on intelligent robots and systems (IROS)* (pp. 450–455). https://doi.org/10.1109/IROS.2017.8202192

Lee, H., Jang, Y., Choe, J. K., Lee, S., Song, H., Lee, J. P., Lone, N., & Kim, J. (2020). 3D-printed programmable tensegrity for soft robotics. *Science Robotics, 5*. https://doi.org/10.1126/scirobotics.aay9024

Levin, S., Lowell de Solórzano, S., & Scarr, G. (2017). The significance of closed kinematic chains to biological movement and dynamic stability. *Journal of Bodywork and Movement Therapies, 21*(3), 664–672. https://doi.org/10.1016/j.jbmt.2017.03.012

Liu, K., Wu, J., Paulino, G. H., & Jerry Qi, H. (2017). Programmable deployment of tensegrity structures by stimulus-responsive polymers. *Scientific Reports, 7*(3511), 2017. https://doi.org/10.1038/s41598-017-03412-6

Ma, S., Chen, M., & Skelton, R. E. (2022). Dynamics and control of clustered tensegrity systems. *Engineering Structures, 264*, 114391. https://doi.org/10.1016/j.engstruct.2022.114391

Maceri, F., Marino, M., & Vairo, G. (2014). An operative algebraic formulation for the unilaterally-constrained mechanical problem of smart tensegrities. *International Journal of Solids and Structures, 51*(19), 3333–3349. ISSN 0020-7683. https://doi.org/10.1016/j.ijsolstr.2014.05.021

Micheletti, A. (2003). The indeterminacy condition for tensegrity towers, a kinematic approach. *Revue Française de Génie Civil, 7*, 329–342. https://doi.org/10.1080/12795119.2003.9692498

Micheletti, A. (2013). Bistable regimes in an elastic tensegrity system. *Proceedings of the Royal Society A: Mathematical, Physical and Engineering Sciences, 469*(2154), 20130052. https://doi.org/10.1098/rspa.2013.0052

Micheletti, A., & Cadoni, D. (2011). Design of single-layer floating-compression tensegrities. In *Proceedings of CSMA 2011, 10e Colloque National en Calcul des Structures, 9-13 May 2011, Giens, France, 2011*.

Micheletti, A., & Podio-Guidugli, P. (2022). Seventy years of tensegrities (and counting). *Archive of Applied Mechanics, 92*, 2525–2548. https://doi.org/10.1007/s00419-022-02192-4

Micheletti, A., & Williams, W. O. (2007). A marching procedure for form-finding for tensegrity structures. *Journal of Mechanics of Materials and Structures, 2*(5), 857–882. https://doi.org/10.2140/jomms.2007.2.857

Micheletti, A., dos Santos, F. A., & Sittner, P. (2018). Superelastic tensegrities: matrix formulation and antagonistic actuation. *Smart Materials and Structures, 27*(10), 105028. https://doi.org/10.1088/1361-665x/aadaa5

Micheletti, A., Ruscica, G., & Fraternali, F. (2019). On the compact wave dynamics of tensegrity beams in multiple dimensions. *Nonlinear Dynamics, 98*, 2737–2753. https://doi.org/10.1007/s11071-019-04986-8

Micheletti, A., dos Santos, F. A., & Guest, S. D. (2023). Prestrain-induced bistability in the design of tensegrity units for mechanical metamaterials. *Applied Physics Letters, 123*(12), 121702. https://doi.org/10.1063/5.0160023

Motro, R. (1984). Forms and forces in tensegrity systems. In *Proceedings of 3rd international conference on space structures. Amsterdam, The Netherlands*.

Motro, R. (1992). Tensegrity systems: The state of the art. *International Journal of Space Structures, 7*(2), 75–83. https://doi.org/10.1177/026635119200700201

Motro, R. (2003). *Tensegrity: structural systems for the future*. London, U.K.: Kogan Page Science.

Oh, C. L., Choong, K. K., Nishimura, T., & Kim, J.-Y. (2022). Multi-directional shape change analysis of biotensegrity model mimicking human spine curvature. *Applied Sciences, 12*, 2377. https://doi.org/10.3390/app12052377

Oppenheim, I. J., & Williams, W. O. (2000). Geometric effects in an elastic tensegrity structure. *Journal of Elasticity, 59*(1–3), 51–65. https://doi.org/10.1023/A:1011092811824

Oppenheim, I. J., & Williams, W. O. (2001). Vibration of an elastic tensegrity structure. *European Journal of Mechanics A/ Solids, 20*, 1023–1031. https://doi.org/10.1016/S0997-7538(01)01181-0

Paul, C., Lipson, H., & Valero Cuevas, F. J. (2005). Design and control of tensegrity robots for locomotion. *IEEE Transactions on Robotics, 22*(5), 944–957. https://doi.org/10.1109/TRO.2006.878980

Pellegrino, S. (1990). Analysis of prestressed mechanisms. *International Journal of Solids and Structures, 26*(12), 1329–1350. https://doi.org/10.1016/0020-7683(90)90082-7

Pellegrino, S. (1993). Structural computations with the singular value decomposition of the equilibrium matrix. *International Journal of Solids and Structures, 30*, 3025–3035. https://doi.org/10.1016/0020-7683(93)90210-X

Pellegrino, S., & Calladine, C. R. (1986). Matrix analysis of statically and kinematically indeterminate frameworks. *International Journal of Solids and Structures, 22*, 409–428. https://doi.org/10.1016/0020-7683(86)90014-4

Quirant, J. (2007). Selfstressed systems comprising elements with unilateral rigidity: Selfstress states, mechanisms and tension setting. *International Journal of Space Structures, 22*, 203–214.

Rhode-Barbarigos, L., Ali, Bel Hadj, & N., Motro, R., & Smith, I. F. C. (2012). Design aspects of a deployable tensegrity-hollow-rope footbridge. *International Journal of Space Structures, 27*, 81–96. https://doi.org/10.1260/0266-3511.27.2-3.81

Rimoli, J. J., & Pal, R. K. (2017). Mechanical response of 3-dimensional tensegrity lattices. *Composites Part B: Engineering, 115*, 30–42. https://doi.org/10.1016/j.compositesb.2016.10.046

Roth, B., & Whiteley, W. (1981). Tensegrity frameworks. *Transactions of the American Mathematical Society, 265*, 419–446. https://doi.org/10.1090/S0002-9947-1981-0610958-6

Sakantamis, K., & Popovic Larsen, O. (2004). The cocoon method: a physical modelling tool for designing tensegrity systems. *International Journal of Space Structures, 19*, 11–20. https://doi.org/10.1260/026635104322988335

Skelton, R. E. (1997). Deployable tendon-controlled structure. U.S. Patent No. 5,642,590, July 1997.

Skelton, R. E., & de Oliveira, M. C. (2009). *Tensegrity systems*. Boston, MA, US: Springer. https://doi.org/10.1007/978-0-387-74242-7

Skelton, R. E., Helton, J. W., Adhikari, R., Pinaud, J. P., & Chan, W. (2002). The mechanical systems design handbook: Modeling, measurement and control. In *Chapter An introduction to the mechanics of tensegrity structures*. London, UK: CRC Press.

Sultan, C. (2009). Tensegrity: 60 years of art, science, and engineering. In Advances in applied mechanics. Advances in applied mechanics (Vol. 43, pp. 69–145). Elsevier. https://doi.org/10.1016/S0065-2156(09)43002-3

Sultan, C., & Skelton, R. E. (1998). Tendon control deployment of tensegrity structures. In V. V. Varadan (Ed.), *Proceedings of SPIE - Volume 3323, smart structures and materials 1998: Mathematics and control in smart structures* (pp. 455–466).

Sultan, C., & Skelton, R. E. (2003). Deployment of tensegrity structures. *International Journal of Solids and Structures, 40*, 4637–4657. https://doi.org/10.1016/S0020-7683(03)00267-1

Tibert, A. G. (2008). Bending-stiff tensegrity masts: Do they exist? In 6th international conference on computation of shell and spatial structures IASS-IACM2008. Ithaca, NY, USA. 28–31 May 2008.

Tran, H. C., & Lee, J. (2010). Advanced form-finding of tensegrity structures. *Computers and Structures, 88*, 237–246. https://doi.org/10.1016/j.compstruc.2009.10.006

Vangelatos, Z., Micheletti, A., Grigoropoulos, C. P., & Fraternali, F. (2020). Design and testing of bistable lattices with tensegrity architecture and nanoscale features fabricated by multiphoton lithography. *Nanomaterials, 10*(4), 652. https://doi.org/10.3390/nano10040652

Wang, B. (2004). *Free-standing tension structures: From tensegrity systems to cable-strut systems*. CRC Press. https://doi.org/10.1201/9781482288308

Xu, X., & Luo, Y. (2010). Multistable tensegrity structures. *Journal of Structural Engineering, 137*(1), 117–123. https://doi.org/10.1061/(ASCE)ST.1943-541X.0000281

Zhang, J. Y., & Ohsaki, M. (2006). Adaptive force density method for form-finding problem of tensegrity structures. *International Journal of Solids and Structures, 43*, 5658–5673. https://doi.org/10.1016/j.ijsolstr.2005.10.011

Zhang, J. Y., & Ohsaki, M. (2007). Stability conditions for tensegrity structures. *International Journal of Solids and Structures, 44*, 3875–3886. https://doi.org/10.1016/j.ijsolstr.2006.10.027

Chapter 2
Minimal Mass Design of Tensegrity Systems

Fernando Fraternali and Gerardo Carpentieri

Abstract This Chapter illustrates a procedure for the minimal mass design of tensegrity systems under yielding and buckling constraints. An optimization method based on an iterative linear programming algorithm is employed, drawing from the multifaceted studies conducted by Robert Skelton and co-workers in the relevant research domain. Minimal mass designs for simply-supported and cantilever beams with tensegrity architecture are shown to exhibit a significantly lower mass, as compared to design procedures based on conventional structural shapes. Newly designed, spider-shaped tensegrity systems are examined for the case of a simply-supported beam, while a cantilever beam example employs the well-known Michell truss topology.

Keywords Tensegrity beams · Minimal mass · Linear programming · Yielding · Buckling

2.1 Introduction

It is known that many engineering designs of tensegrity structures draw inspiration from natural shapes. As extensively shown in other Chapters of this collection, tensegrity concepts are evident in various natural phenomena, including cellular structures, the locomotion systems of humans and animals, and the microarchitecture of spider silk threads, among other examples (refer also to Ingber 1997; Skelton & de Oliveira 2009 and references therein). The presence of tensegrity concepts in natural shapes is explained by the ability of tensegrity systems to deliver remarkably high levels of stiffness and strength in lightweight structures, due to the absence of bending stresses (Skelton & de Oliveira, 2009). In the field of structural engineering, material savings are at the top of the global agenda due to the urgent need to reduce carbon dioxide

F. Fraternali (✉) · G. Carpentieri
Department of Civil Engineering, University of Salerno, Via Giovanni Paolo II 132, 84084 Fisciano, SA, Italy
e-mail: f.fraternali@unisa.it

G. Carpentieri
e-mail: gcarpentieri@unisa.it

F. Fraternali and J. J. Rimoli (eds.), *Tensegrity Systems*, CISM International Centre for Mechanical Sciences 2, https://doi.org/10.1007/978-3-031-82283-4_2

emissions and the rising costs of raw materials (Pomponi et al., 2020). Similarly, in aeronautical and aerospace engineering, the emphasis is on the design lightweight and deployable mechanical systems. This priority aims to mitigate the impact of gravity and facilitate remote control of the structure (Miura & Pellegrino, 2020).

The present Chapter illustrates a minimal mass design procedure for tensegrity beams. The presented algorithm is inspired by the seminal work conducted in this area of research by Robert Skelton and co-workers over recent years (Nagase & Skelton, 2014; Skelton et al., 2014; Chen & Skelton, 2020). It recursively employs a linear programming procedure, by handling both yielding and buckling constraints (Sect. 2.3). The minimal mass design of simply-supported and cantilever beams with tensegrity architecture is illustrated in Sect. 2.4, where comparisons with design procedures based on conventional structural shapes are given. We end with concluding remarks and directions for future research in Sect. 2.5.

2.2 Employed Notation

Key notations used throughout this Chapter are summarized below for the reader's convenience. The symbols $\boldsymbol{x}$, $\boldsymbol{w}$, and $\boldsymbol{A}$ hereafter represent, respectively, the vector of force densities in the strings and bars of a tensegrity structure, the vector of nodal forces, and the corresponding static (or equilibrium) matrix (or operator) of the system. For alternative notations of these variables, please refer to the Chapter on basic concepts.

2.3 A Linear Programming Algorithm for the Minimal Mass Design of Tensegrity Systems

We summarize hereafter an iterative linear programming procedure extensively presented in Nagase & Skelton (2014), Skelton et al. (2014), Chen & Skelton (2020) for the minimal mass design of tensegrity systems. Let us refer to the tensegrity beam shown in Fig. 2.1, which will be diffusely discussed in the following Sect. 2.4.1. Throughout the Chapter, we let h denote the maximum height of a tensegrity profile (cf. Fig. 2.1).

For the sake of simplicity, we assume that both the n_b members carrying compression forces (bars) and the n_s tensile members (cables or strings), which form the structure, are made of the same material with mass density ρ, Young modulus E and yielding stress $\bar{\sigma}_Y$. Assuming *full stress* loading conditions, we write the yielding constraints of the strings into the following form

$$\sigma_{s_i} = \bar{\sigma}_Y, \quad i = 1, ..., n_s \tag{2.1}$$

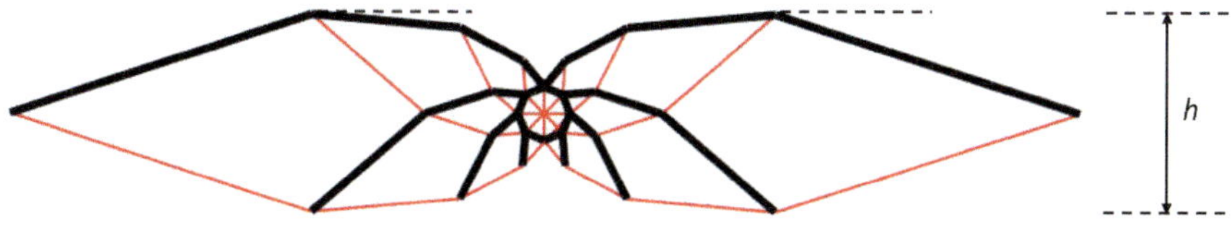

Fig. 2.1 Illustrative example of a tensegrity beam with bar elements represented by black/thick lines and strings depicted by red/thin lines

σ_{s_i} denoting the design value of the axial stress carried by the i-th string. We design the bars under yielding and (local) buckling constraints. The latter require that the axial force carried by the generic bar, say the i-the one, does not exceed the Eulerian critical load corresponding to hinged boundary conditions: $N_{b_i}^{cr} = \pi^2 E I_{b_i} / b_i^2$. Here I_{b_i} is the (minimum) moment of inertia of the cross-section, and b_i is the bar length. Assuming that the generic bar exhibits circular cross section with radius r_{b_i}, area $A_{b_i} = \pi r_{b_i}^2$ and moment of inertia $I_{b_i} = \pi r_{b_i}^4 / 4$, the Eulerian buckling stress $\bar{\sigma}_{B_i} = N_{b_i}^{cr} / A_{b_i}$ is easily computed as $\bar{\sigma}_{B_i} = \frac{\pi^2 E}{4} \left(\frac{r_{b_i}}{b_i} \right)^2 = \frac{\pi E}{4} \frac{A_{b_i}}{b_i^2}$. Under the assumption of full stress loading conditions also in the bars, we write the combined yielding-buckling constraints of such elements into the form

$$\sigma_{b_i} = \min \left(\bar{\sigma}_Y, \frac{\pi E}{4} \frac{A_{b_i}}{b_i^2} \right), \quad i = 1, ..., n_b \tag{2.2}$$

where σ_{b_i} is the design value of the normal stresses carried by the i-th bar.

We now introduce the vector $\boldsymbol{x}$ of the *force densities* in the bars and strings, which are defined as the ratios between the axial forces t_{b_i} and t_{s_i} carried by such members and the corresponding lengths b_i and s_i. We assume that such quantities are positive in compression for the bars, and positive in tension for the strings. The overall mass of the structure is easily computed as follows

$$m = \boldsymbol{c}^T \boldsymbol{x} \tag{2.3}$$

T denoting the transposition symbol, and it results in

$$\boldsymbol{c} = \left(\frac{\rho b_1^2}{\sigma_{b_1}} \dots \frac{\rho b_{n_b}^2}{\sigma_{b n_b}} \mid \frac{\rho s_1^2}{\sigma_{s_1}} \dots \frac{\rho s_{n_s}^2}{\sigma_{s n_s}} \right)^T \tag{2.4}$$

The equilibrium equations of the structure require

$$\boldsymbol{A}\boldsymbol{x} = \boldsymbol{w} \tag{2.5}$$

where, as mentioned earlier, $\boldsymbol{A}$ represents the *static matrix*, and $\boldsymbol{w}$ denotes the *nodal force vector* of the system (Nagase & Skelton, 2014). Based on these preliminaries, we now introduce the following *linear programming* algorithm for designing a minimal mass tensegrity structure

$$\begin{aligned} &\underset{\boldsymbol{x}}{\text{minimize}} \quad m = \boldsymbol{c}^T \boldsymbol{x} \\ &\text{subject to} \quad \begin{cases} \boldsymbol{A}\boldsymbol{x} = \boldsymbol{w} \\ \boldsymbol{x} \geq 0. \end{cases} \end{aligned} \tag{2.6}$$

The linear programming problem (2.6) needs to be solved in iterative form, in order to account for the combined buckling-yielding constraints of the bars (2.2). Let $m^{'}$ denote the minimal mass obtained by solving such a problem with $\sigma_{b_i} = \bar{\sigma}_Y$ in the bars. The post-processing of the solution of problem (2.6) drives us to compute the axial forces $t_{b_i}^{'}$ carried by the bars and the axial forces $t_{s_i}^{'}$ carried by the strings. The enforcement of the constraints (2.2) gives the following, updated values of the cross-section areas of the bars

$$A_{b_i}^{''} = \begin{cases} \sqrt{t_{b_i}^{'}/(\pi E/4b_i^2)} & (t_{b_i}^{'} \leq t_{b_i}^{*}) \\ t_{b_i}^{'}/\bar{\sigma}_Y & (t_{b_i}^{'} > t_{b_i}^{*}) \end{cases} \tag{2.7}$$

with $t_{b_i}^{*} = (\bar{\sigma}_Y)^2/(\pi E/4b_i^2)$ (cf. Nagase & Skelton 2014). The redesign of the bars finally leads us to compute the following updated mass of the structure

$$m^{''} = \sum_{i=1}^{n_b} \rho\, A_{b_i}^{''}\, b_i \;+\; \sum_{i=1}^{n_s} \frac{\rho s_i^2}{\sigma_{s_i}}\, x_{s_i}^{'} \tag{2.8}$$

The mass minimization procedure can be arrested when it results in $|(m^{''} - m^{'})/m^{'}| \leq$ tol, tol denoting a prescribed tolerance. If, conversely, one obtains $|(m^{''} - m^{'})/m^{'}| >$ tol at the current iteration, problem (2.6) needs to be solved again, assuming $\sigma_{b_i} = t_{b_i}^{'}/A_{b_i}^{''}$. The execution of a correction step, based on Eqs. (2.7) and (2.8), will next tell us if should continue iterating the above predictor-corrector strategy or not. A Matlab®script can easily implements such a procedure, see, e.g., the 'tensopt' code given in Fraternali & Carpentieri (2015).

2.4 Minimal Mass Tensegrity Beams

The present section is aimed at comparing conventional design procedures of simply-supported and cantilever beams, which are based on the adoption of I-shaped profiles for the cross-section (Fig. 2.2), with alternative designs of tensegrity shapes obtained through the minimal mass procedure presented in the previous section.

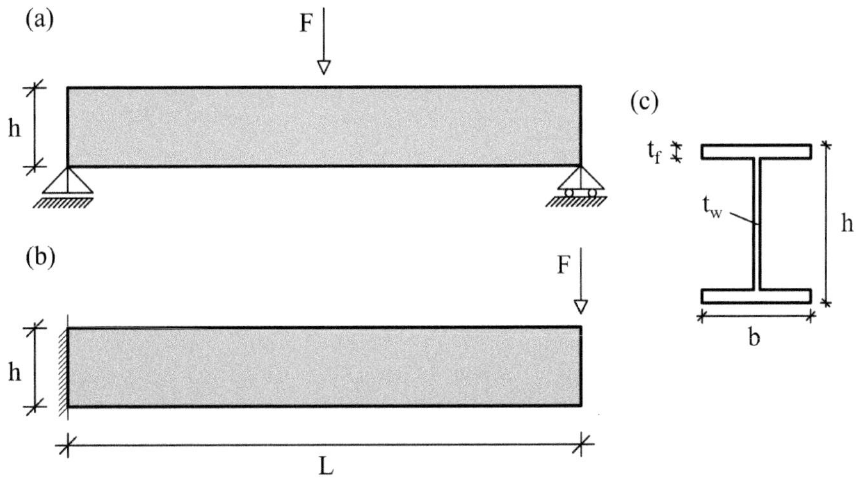

Fig. 2.2 Solid beams loaded by the self-weight and a concentrated force. **a** simply-supported beam; **b** clamped beam; **c** I-shaped cross section

2.4.1 Simply-Supported Beams

Let consider first the simply-supported beam shown in Fig. 2.2a, which is equipped with the I-shaped cross section depicted in Fig. 2.2c. The area A and the moment of inertia I of the cross section (with respect to the horizontal axis passing through the center of mass) are respectively given by

$$A = bh - (b - t_w)\left(h - 2t_f\right), \qquad I = \frac{bh^3}{12} - \frac{(b - t_w)\left(h - 2t_f\right)^3}{12} \tag{2.9}$$

The maximum bending moment at the mid-span and the maximum shear force at the extremity are computed through

$$M = \frac{FL}{4} + \frac{\rho g A L^2}{8} \tag{2.10}$$

$$T = \frac{F}{2} + \frac{\rho g A L}{2} \tag{2.11}$$

where ρ is the mass density of the material of the beam and $g = 9,81 m/s^2$ if the gravity. Then, one can write the yielding constraint of the material into the following form

$$\sigma = \max\left(\sigma_{id}^{(1)}, \sigma_{id}^{(2)}\right) \leq \frac{\sigma_y}{s} \tag{2.12}$$

Table 2.1 Material properties

Steel		Aluminum	
$\rho\ [kg/m^3]$	7862	$\rho\ [kg/m^3]$	2700
$\sigma\ [N/m^2]$	2.35×10^8	$\sigma\ [N/m^2]$	1.3×10^8
$E\ [N/m^2]$	206×10^9	$E\ [N/m^2]$	70×10^9
s	1.05	s	1.10

Here, s is a safety factor, $\sigma_{id}^{(1)}$ is the maximum stress of the mid-span cross sectional (due to the bending moment) and $\sigma_{id}^{(2)}$ is the maximum stress of the extremities cross sectional (due to the shear forces). The above maximum stresses can be computed with the following relationships

$$\sigma_{id}^{(1)} = \frac{M}{I}\left(\frac{h}{2}\right) \tag{2.13}$$

$$\sigma_{id}^{(2)} = \sqrt{\alpha}\left(\frac{TS}{t_w I}\right) \tag{2.14}$$

where $\alpha = 3$ for the Von Mises yielding failure criterion, $\alpha = 4$ for the Tresca yielding failure criterion, and S is the static moment of area given by (Beer et al., 2014)

$$S = bt_f\frac{\left(h - t_f\right)}{2} + \frac{t_w}{2}\left(\frac{h}{2} - t_f\right)^2 \tag{2.15}$$

By assuming $\alpha = 3, t_f = h/20, b = h/2, t_w = h/25$, we obtain the design charts shown in Fig. 2.3 for steel and aluminum beams. Such charts plot the overall mass m of the beam and the cross section height h against the amplitude of the applied force F, which are obtained making use of the design criterion (2.12), for different values of the beam length L. The assumed material properties are listed in Table 2.1. As a case study, let's consider a steel beam with $L = 5$ m, $F = 250$ kN. The design chart shown in Fig. 2.3 returns for such a beam an overall mass of 467.016 kg.

We now formulate a tensegrity design of s simply-supported beam, which is inspired by the topology of the well-known Michell truss problem (refer, e.g., to Michell 1904; Skelton & de Oliveira 2009, 2010). Our tensegrity design employs two mirrored Michell trusses $\mathcal{M}_1$ and $\mathcal{M}_2$ with height h, each covering one half of the span, so as to for a beam that mimics the shape of a sector of a spider orb-web (Sanders et al., 2020). We will refer to such a system as a *spider beam* in the following. Let us refer to one of the trusses that compose the spider beam, say, e.g., $\mathcal{M}_1$. It is formed by nodes lying on $q + 1$ circles of radii $r_i = ar_{i-1}$, with $i = 0, ..., q$, centered at the support point, which are spaced by a given angle ϕ. The nodes are joined by spiral members of lengths cr_i that form an angle β with the radial segment joining the support point with one end of the member. Here ϕ and a denote design

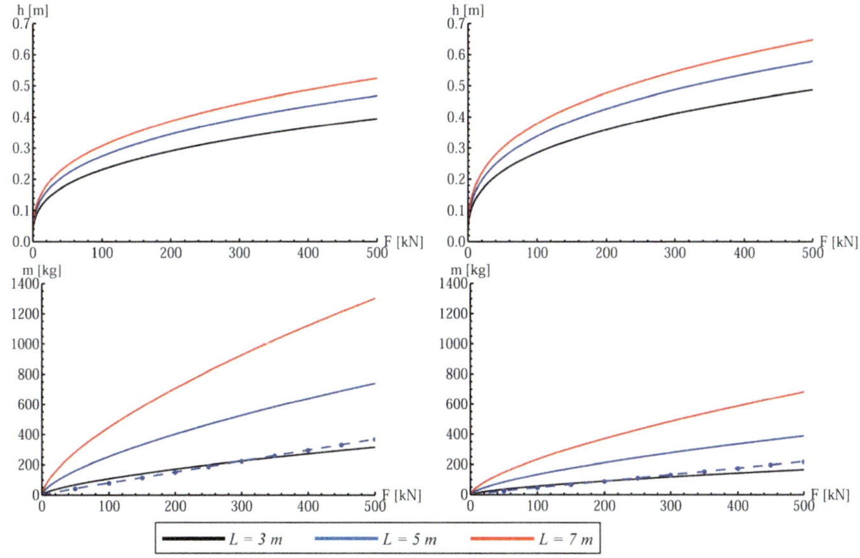

Fig. 2.3 Design charts for the cross section height (h) and the overall mass (m) of steel (left) and aluminum (right) simply-supported beams of various lengths L, which are loaded by the self-weight and a concentrated force F at the mid span. The $m - F$ plots include tensegrity solutions (dashed curves), which were obtained for $q = 7$ and two values of the height h (cf. Fig. 2.1): $h = 0, 371677$ m (left), and $h = 0, 459178$ m (right)

parameters, and q indicates the *complexity* of the system and it results in (see Skelton & de Oliveira 2009, Chap. 4, for further details)

$$a = \frac{\sin\beta}{\sin(\beta+\phi)}, \qquad c = \frac{\sin\phi}{\sin(\beta+\phi)} \tag{2.16}$$

Equation (2.16) can be solved for β and c in terms of a and ϕ, obtaining

$$\beta = \cos^{-1}\left(\frac{1 - a\cos(\phi)}{\sqrt{a^2 - 2a\cos(\phi) + 1}}\right), \qquad c = \sqrt{a^2 - 2a\cos(\phi) + 1} \tag{2.17}$$

For a given value of the transverse force F shown in Fig. 2.2a, each half of the beam is loaded by a tip force with intensity $F/2$. The axial forces in each member of $\mathcal{M}_1$ and $\mathcal{M}_2$ can be obtained as illustrated in Skelton & de Oliveira (2009) (Chap. 4), and the minimal mass design of the structure can be carried out employing the linear programming algorithm illustrated in Sect. 2.3. Gravity in the downward direction is added. Let us assume now $a = (h/L)^{1/q}$, $\phi = \pi/q$ and $r_q = h/2$. Figure 2.4 shows the minimal mass spider beams obtained for complexities ranging from $q = 1$ to 10, when considering bars and strings made of steel.

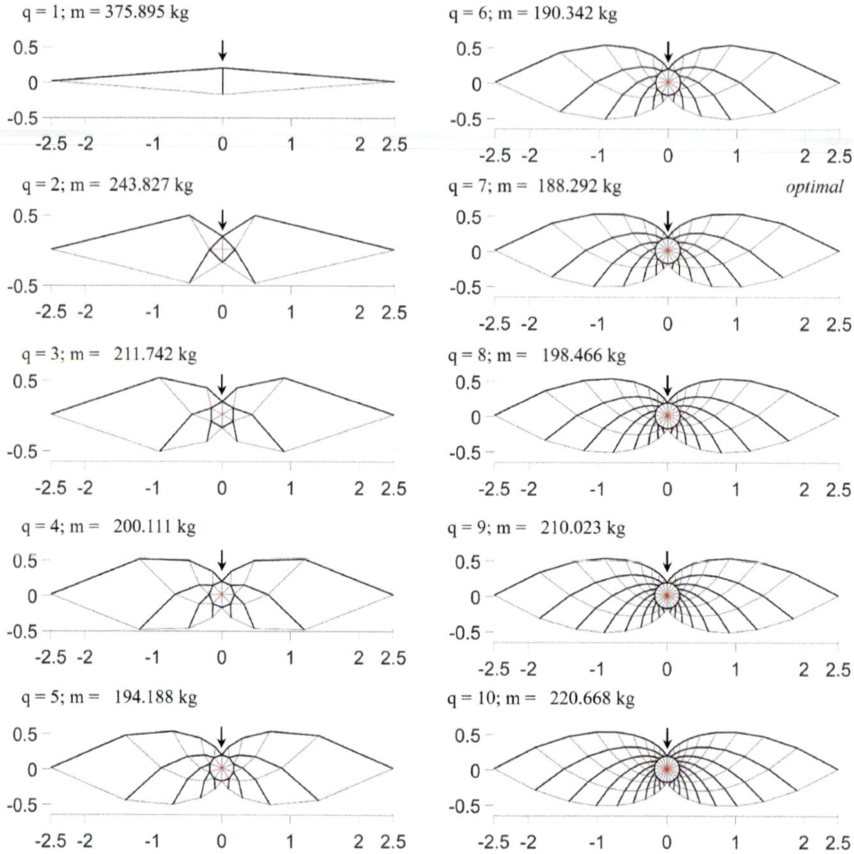

Fig. 2.4 Minimal mass designs of simply-supported spider beams made of steel for different complexities q ($F = 250$ kN)

For $F = 250$ kN, we obtain masses ranging from 375.895 kg for $q = 1$ to 188.292 kg for $q = 7$ (optimal complexity). It is seen that the tensegrity design of the case with $q = 7$ exhibits a mass that is 2.48 times lower than the mass of the corresponding I-shaped beam (cf. Fig. 2.3). In the same case ($q = 7$), the tensegrity minimal design based on the algorithm presented in Sect. 2.3 returns a system with radii of the bars varying from 1.7 mm to 42.7 mm, and radii of the strings varying from 7.8 mm to 28.5 mm.

2.4.2 Cantilever Beams

We now move on to consider the cantilever beam shown in Fig. 2.2b with the cross section illustrated in Fig. 2.2c. The area and moment of inertia of such section are given by Eq. (2.9). The maximum bending moment and the maximum shear force under the action of a tip force with intensity F and the self-weight is reached at the clamped support, being equal to

$$M = FL + \frac{\rho g A L^2}{2} \tag{2.18}$$

$$T = F + \rho g A L \tag{2.19}$$

The yielding condition of the material is given by

$$\sigma_{id} = \sqrt{\sigma^2 + \alpha \tau^2} \leq \frac{\sigma_y}{s} \tag{2.20}$$

where:

$$\sigma = \frac{M}{I}\left(\frac{h}{2}\right) \tag{2.21}$$

$$\tau = \frac{TS}{t_w I} \tag{2.22}$$

Supposing $\alpha = 3$, $t_f = h/20$, $b = h/2$, $t_w = h/25$, as in the case of the simply-supported beam, we obtain the design charts illustrated in Fig. 2.5. It is worth noting that, for $L = 3$ m, $F = 100$ kN, we obtain a steel I-shaped beam with 0.36897 m height and a total mass of 276.143 kg.

We now consider tensegrity designs of a cantilever beam that employ the Michell topology illustrated in the previous section over the entire span of the beam. We assume $a = [h/(2L)]^{1/q}$, $\phi = \pi/q$, $r_q = h/2$ and $F = 100$ kN, to obtain the minimal mass designs shown in Fig. 2.6 for complexities varying from $q = 1$ to 10 and bars and strings made of steel. The masses of the tensegrity cantilever beams illustrated in Fig. 2.6 range from 189.638 kg ($q = 1$) to 33.6575 kg ($q = 10$). For $q = 10$ (local optimum among the examined complexities) one observes that the tensegrity beam exhibits a mass that is 8.2 times lower that the mass of the I-shaped cantilever beam (cf. Fig. 2.5). It is worth noting that the tensegrity minimal design returns, in such a case ($q = 10$), a system with radii of the bars varying from 4.7 mm to 15.1 mm, and radii of the strings varying from 3.3 mm to 12.2 mm.

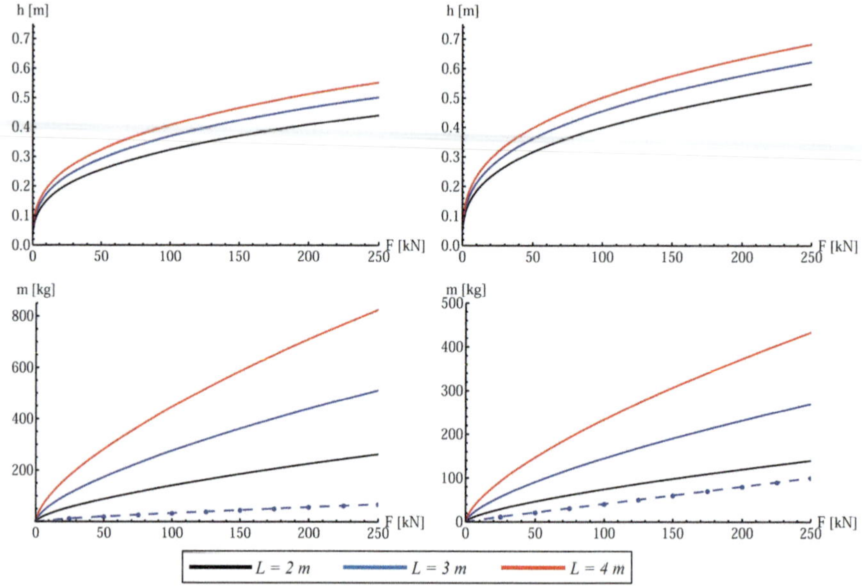

Fig. 2.5 Design charts for the cross section height (h) and the overall mass (m) of steel (left) and aluminum (right) cantilever beams of various lengths L, which are loaded by the self-weight and a concentrated force F at the tip. Dashed curves shows tensegrity solutions for $q = 10$ and the following two values of h: $h = 0{,}36897$ m (left), and $h = 0{,}456636$ m (right)

2.5 Concluding Remarks

We have illustrated minimal mass designs of simply-supported and cantilever beams with tensegrity architecture, which have been obtained though an iterative linear programming algorithm handling yielding and buckling constraints. A comparative study has shown that tensegrity beams are able to exhibits masses that are nearly 100 times lower than those of solid beams with I-shaped cross sections made of the same material. Such remarkable mass savings are very useful for the design of lightweight and sustainable structures to be employed for a variety of engineering applications (Skelton & de Oliveira, 2009; Miura & Pellegrino, 2020; Seixas et al., 2021). Future extensions of the present study may examine minimal mass designs of tensegrity structures that cover topology, geometry and members' sizing optimizations. Additional lines of future research may be concerned with optimal design strategies that examine different objective functions, such as, e.g., minimum or maximum stiffness and/or frequency bandgap amplitudes (see the related Chapters of the present collection), with the aim of designing of novel, bionspired mechanical metamaterials with tensegrity architecture. The minimal mass procedure presented in this Chapter could also be applied to the optimal design of no-tension structures, potentially in conjunction with the techniques discussed in Amendola et al. (2023). It could be used

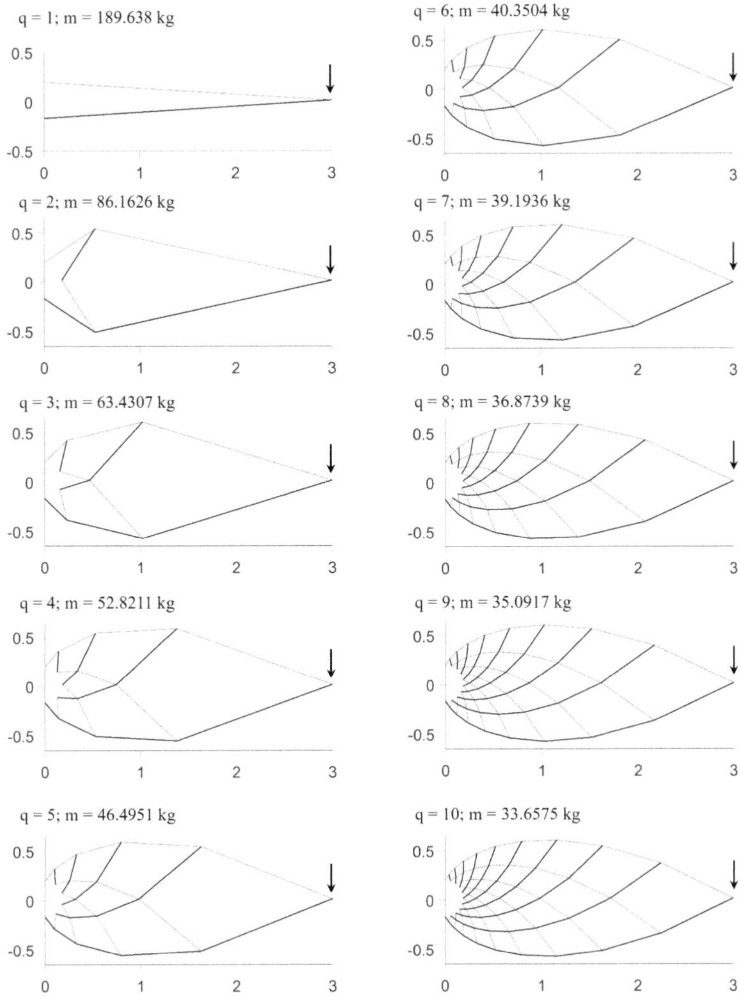

Fig. 2.6 Minimal mass designs of tensegrity cantilever beams made of steel for different complexities q ($F = 100$ kN)

for the design of tensegrity braces for anti-seismic structures (Santos et al., 2022), origami systems and bio-inspired materials (Fraternali et al., 2020, 2024a, b).

References

Amendola, A., Mattei, O., Milton, G. W., & Seppecher, P. (2023). The obstacle problem in no-tension structures and cable nets. *Proceedings of the Royal Society A, 479*(2269), 20220229.

Beer, F., Johnston, E., & DeWolf, J. (2014). *Mechanics of materials*, 7th edn. McGraw-Hill.

Chen, Muhao, & Skelton, Robert E. (2020). A general approach to minimal mass tensegrity. *Composite Structures, 248*, 112454.

Fraternali, F., Carpentieri, G. (2015). Tensopt code. https://ww.fernandofraternaliresearch.com

Fraternali, F., Stehling, N., Amendola, A., Tiban Anrango, B. A., Holland, C., & Rodenburg, C. (2020). Tensegrity modelling and the high toughness of spider dragline silk. *Nanomaterials, 10*(8), 1510.

Fraternali, F., de Castro Motta, J., Germano, G., Babilio, E., & Amendola, A. (2024a). Mechanical response of tensegrity-origami solar modules. *Applications in Engineering Science, 17*, 100174.

Fraternali, F., Babilio, E., Nazifi Charandabi, R., Germano, G., Luciano, R., & Spagnuolo, G. (2024b). Dynamic origami solar eyes with tensegrity architecture for energy harvesting Mashrabiyas. *Applications in Engineering Science, 19*, 100190.

Ingber, D. E. (1997). Tensegrity: the architectural basis of cellular mechanotransduction. *Annual Review of Physiology, 59*(1), 575–599.

Michell, A. G. M. (1904). Lviii. The limits of economy of material in frame-structures. *The London, Edinburgh, and Dublin Philosophical Magazine and Journal of Science, 8*(47), 589–597.

Miura, K., & Pellegrino, S. (2020). *Forms and concepts for lightweight structures*. Cambridge University Press.

Nagase, K., & Skelton, R. E. (2014). Minimal mass tensegrity structures. *Journal of The International Association for Shell and Spatial Structures, 55*(1), 37–48.

Pomponi, F., Hart, J., Arehart, J. H., D'Amico, B. (2020). Buildings as a global carbon sink? a reality check on feasibility limits. *One Earth, 3*(2), 157–161.

Sanders, E. D., Ramos, A. S., & Paulino, G. H. (2020). Topology optimization of tension-only cable nets under finite deformations. *Structural and Multidisciplinary Optimization, 62*, 559–579.

Santos, F. A., Caroço, C., Amendola, A., Miniaci, M., & Fraternali, F. (2022). 3D tensegrity braces with superelastic response for seismic control. *International Journal for Multiscale Computational Engineering, 20*(5), 53–64.

Seixas, M., Moreira, L. E., Stoffel, P., Bina, J., Ripper, J.L.M., Ferreira, J. L., & Ghavami, K. Analysis of a self-supporting bamboo structure with flexible joints. *International Journal of Space Structures, 36*(2), 137–151.

Skelton, R. E., Fraternali, F., Carpentieri, G., & Micheletti, A. (2014). Minimum mass design of tensegrity bridges with parametric architecture and multiscale complexity. *Mechanics Research Communications, 58*, 124–132.

Skelton, R. E., & de Oliveira, M. C. (2009). *Tensegrity systems*. Berlin: Springer.

Skelton, R. E., & de Oliveira, M. C. (2010). Optimal tensegrity structures in bending: The discrete michell truss. *Journal of the Franklin Institute, 347*(1), 257–283.

Chapter 3
Bandgap Wave Dynamics of Tensegrity Lattices

Fernando Fraternali, Julia de Castro Motta, and Luca Placidi

Abstract This Chapter studies the wave dynamics of systems obtained by combining lumped masses with tensegrity units to form one-dimensional mechanical metamaterials. The analyzed tensegrity units behave as elastic springs with mechanical response that can be finely adjusted by applying an initial state of self-stress to the system. One-dimensional chains of tensegrity prisms are shown to exhibit a tunable, frequency bandgap response under small amplitude compression loading in the dynamic regime. Both monoatomic and diatomic mass-spring models are studied, by examining various arrangements of two types of tensegrity prisms and lumped masses. The continuum limits of the analyzed discrete models are also derived, with reference to the small wave number regime, and challenging engineering applications of bandgap metamaterials with tensegrity architecture are discussed.

Keywords Tensegrity lattices · Frequency bandgaps · Monoatomic · Diatomic · Continuum limits

3.1 Bandgap Tensegrity Metamaterials

Acoustic and mechanical metamaterials are architectured composite materials with an optimally designed internal architecture, which exhibit exceptional properties surpassing those of the component materials (Kadic et al., 2019). These systems can

F. Fraternali (✉) · J. de Castro Motta
Department of Civil Engineering, University of Salerno, Via Giovanni Paolo II 132, 84084 Fisciano, SA, Italy
e-mail: f.fraternali@unisa.it

J. de Castro Motta
e-mail: jdecastromotta@unisa.it

L. Placidi
International Telematic University Uninettuno, Corso Vittorio Emanuele II 39, 00186 Rome, Italy
e-mail: luca.placidi@uninettunouniversity.net

F. Fraternali and J. J. Rimoli (eds.), *Tensegrity Systems*, CISM International Centre for Mechanical Sciences 2, https://doi.org/10.1007/978-3-031-82283-4_3

exhibit, e.g., a bandgap frequency response due to Bragg-scattering and internal-resonance phenomena at the wavelength and sub-wavelength scales (Kadic et al., 2019; Liu et al., 2000; Lu et al., 2009).

Metamaterials formed by tessellating tensegrity units in one-, two- and three-dimensions exhibit a peculiar response, which is tunable through internal and external prestress and can vary from stiffening to softening (Amendola et al., 2014; Fraternali et al., 2015; Micheletti et al., 2019). The possibility to finely adjust the stiffness properties of tensegrity units has been used to design linear metamaterials with tunable frequency bandgaps in the literature (Amendola et al., 2018; Krushynska et al., 2018). It has also been shown that geometrically nonlinear tensegrity metamaterials can support the propagation of solitary waves with compact support under impact loading (Fraternali et al., 2012; Micheletti et al., 2019; Amendola, 2023; Fraternali et al., 2014; de Castro Motta et al., 2024), as will be demonstrated in a dedicated chapter of this book.

This Chapter investigates the frequency bandgap response of one- dimensional mass-spring chains with tensegrity architecture by reviewing existing literature and presenting new analytical and numerical findings. This type of response arises from the *Bragg-scattering* phenomenon, which is associated with the destructive interference of mechanical waves radiating from the units of a periodic array of 'scatterers' (Huber, 2018). As a result, harmonic waves are unable to propagate through the discrete system when their frequencies fall within certain *bandgaps* (see Chap. 1 of Huber 2018 and Chap. 22 of Ashcroft and Mermin 1976). Monoatomic and diatomic systems are studied, by considering two different types of tensegrity prism as the elastic members of the mass-springh chain.

A distinctive property of tensegrity bandgap systems is their ability to finely tune frequency bandgaps by adjusting prestress variables, offering additional flexibility to modulate the stiffness of the system's elastic members, beyond the conventional approach of optimizing the size and materials of these components (Amendola et al., 2018).

The novelty of the study presented in this Chapter lies in the modeling of the coupling effects between the axial and twisting motions of tensegrity units, as well as in the development of continuum models for the analyzed systems in the limit where the size of the unit cells approaches zero. This coupling introduces the concept of equivalent masses and allows for further tuning of the frequency bandgap response of the system. The Chapter concludes by exploring the potential applications of tensegrity metamaterials in designing novel acoustic devices with enhanced tunability and outlines future research directions and challenges for extending this area of research.

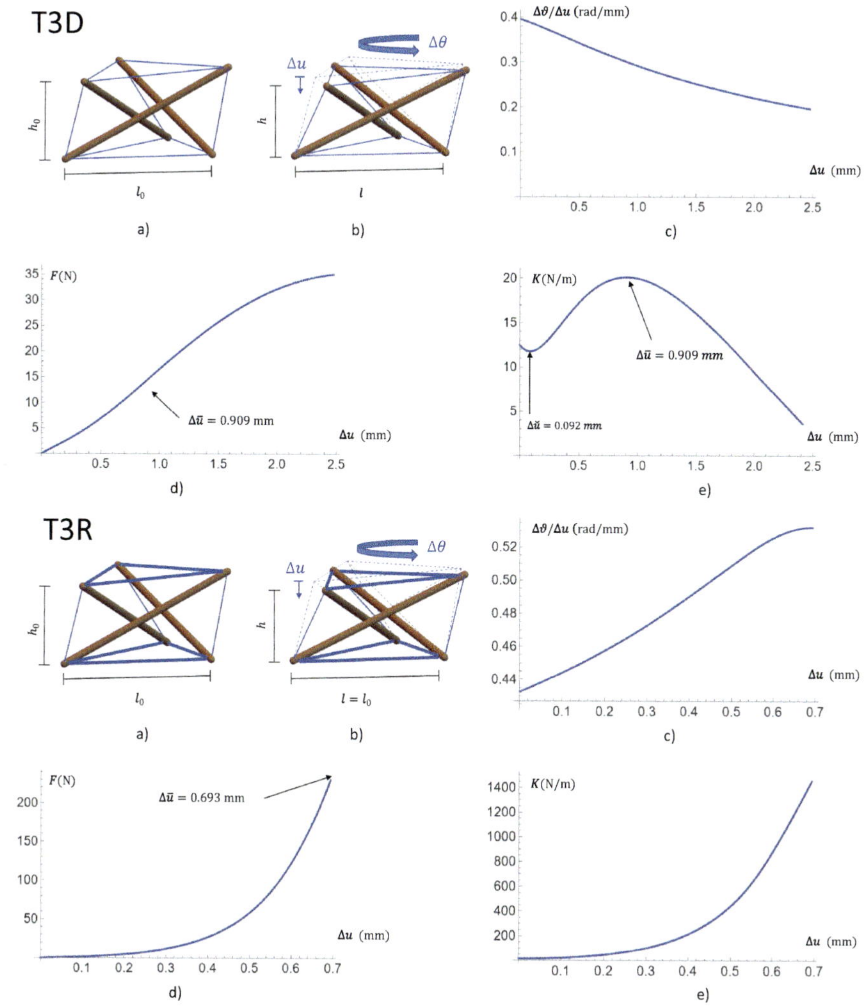

Fig. 3.1 Illustration of the mechanical response of the T3D (top) and T3R (bottom) units. **a** Reference configuration. **b** Deformed configuration. **c** Law relating the relative twist angle $\Delta\vartheta$ between the terminal bases with the relative axial displacement Δu. **d** Axial force F versus relative axial displacement Δu. **e** Tangent axial stiffness K versus relative axial displacement Δu

3.2 Mechanical Response of Tensegrity Units Under Compression Loading

We analyze in this section various assemblies of two types of micro-scale tensegrity prisms: minimal regular prisms equipped with three bars (Skelton and de Oliveira, 2009), which are hereafter referred to as T3 prisms, and $\theta = 1$ prisms (Mascolo et al., 2018), hereafter referred to as Tθ prisms.

With reference to the T3 prisms (Fig. 3.1), we let p_0 denote the initial axial strain (or prestrain) of the cross-strings; and, with reference to the reference configuration subject to p_0 (self-standing configuration), we let h_0, b_0, d_b, ℓ_0, s_0, d_s and ϑ_0 respectively denote the height of the prism, the length of the bars, the diameter of the bars, the length of the base strings, the length of the cross strings, the diameter of the strings, and the initial angle of twist between the terminal bases. We also use the symbols E_b and E_s to indicate the Young moduli of the material of the bars, and the material of the strings, respectively.

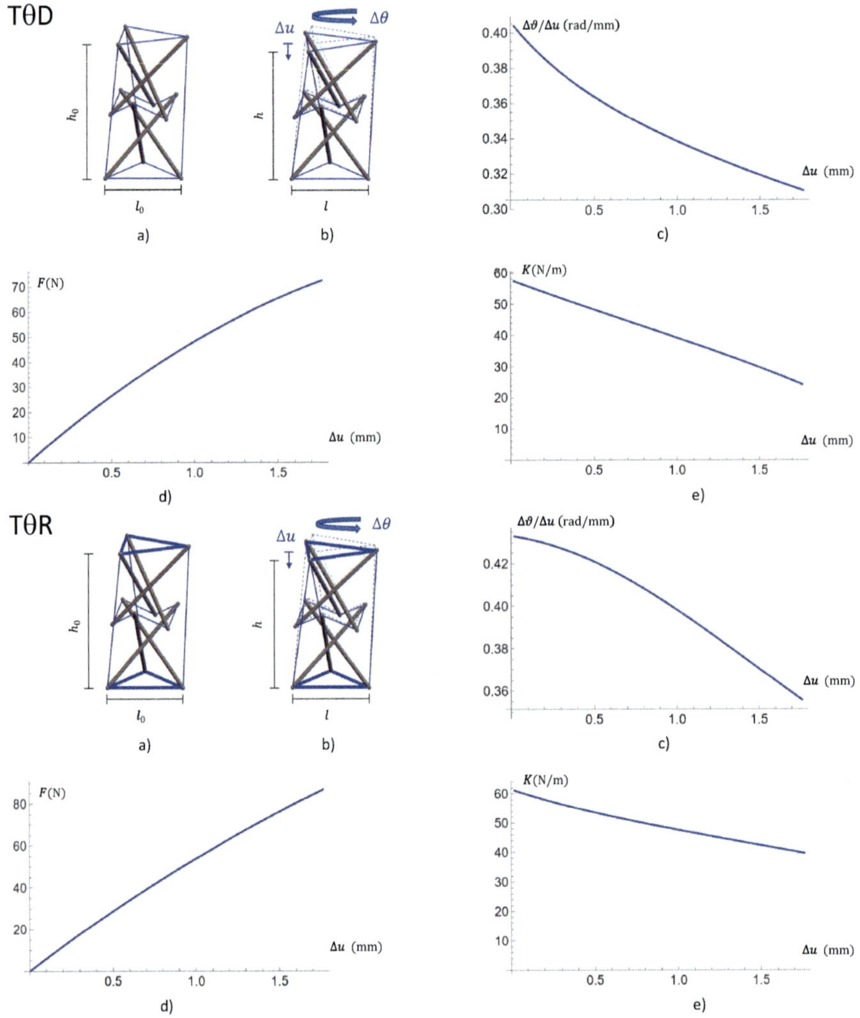

Fig. 3.2 Illustration of the mechanical response of the TθD (top) and TθR (bottom) units. **a** Reference configuration. **b** Deformed configuration. **c** Law relating the relative twist angle $\Delta\vartheta$ between the terminal bases with the relative axial displacement Δu. **d** Axial force F versus relative axial displacement Δu. **e** Tangent axial stiffness K versus relative axial displacement Δu

Table 3.1 Geometrical and mechanical properties of T3 prisms

p_0 %	h_0 (mm)	b_0 (mm)	d_b (mm)	E_b (MPa)	ℓ_0 (mm)	s_0 (mm)	d_s (mm)	E_s (MPa)	ϑ_0 (deg)
1	5.5	11.21	0.80	120	8.71	6.06	0.28	5.48	150

Table 3.2 Geometrical properties of Tθ prisms

p_0 %	h_0 (mm)	b_0 (mm)	d_b (mm)	ℓ_0 (mm)	s_v (mm)	s_c (mm)	d_s (mm)	ϑ_0 (deg)	β_0 (deg)
10	11.0	8.71	0.80	6.10	11.10	3.05	0.28	24.45	16.05

For what concerns the Tθ prisms, we instead let $p_0, h_0, b_0, d_b, \ell_0, s_v, s_c, d_s, \vartheta_0$ and β_0 respectively denote the prestrain of the base strings, the height of the prism, the length of the bars, the diameter of the bars, the length of the base strings, the length of the vertical (or cross strings), the length of the intermediate (or inner) strings, the diameter of the strings, the initial angle of twist between the terminal bases, and the initial angle formed by the intermediate strings with the horizontal plane. The quantities not affected by the subscript '0' refer to the deformed configuration of the prism (Figs. 3.1, 3.2).

Tables 3.1 and 3.2 illustrate the geometrical and mechanical properties of the tensegrity systems under consideration, all of which make use of Ti6Al4V titanium alloy for the bars (Young modulus 120 MPa), and Spectra cables for the strings (Young modulus 5.48 MPa). Such materials can be conveniently employed to 3D print the analyzed units (see Amendola et al. 2015). Figures 3.1 and 3.2 illustrate the response of the analyzed units under uniform compression loading, which is obtained by applying two opposed axial forces F to the centers of mass of the terminal bases of the prisms. Such a loading condition produces a relative longitudinal displacement $\Delta u = h_0 - h$ between the terminal bases, assumed positive in compression, and a relative angle of twist $\Delta\vartheta = \vartheta - \vartheta_0$ between the same elements, assumed positive counterclockwise (refer, e.g., to Fraternali et al. 2015; Mascolo et al. 2018 and references therein).

We study both prisms that are formed by terminal bases free to deform in their own planes (T3D and TθD prisms), and prisms that are such that their bases transform rigidly under the applied loading (T3R and TθR prisms). The latter are aimed at modeling situations in which the bases of the prisms are perfectly bonded to lumped masses, as we will see in the subsequent sections. The prisms with deformable bases will be instead used to form mass-spring chains such that the tensegrity units are free to slide tangentially against the lumped masses.

The results shown in Fig. 3.1-top highlight that the T3D prism initially exhibits an elastic stiffening response (i.e., a slope K of the F versus Δu curve that increases with increasing values of Δu), for axial displacements ranging in the interval $\Delta\tilde{u} < \Delta u < \Delta\bar{u}$ ($\Delta\tilde{u} = 0.092$ mm; $\Delta\bar{u} = 0.909$ mm, cf. Fig. 3.1-top). Such a response is followed by an elastically softening behavior (axial stiffness K decreasing with Δu) for $\Delta u > \Delta\bar{u}$. The $\Delta\vartheta/\Delta u$ versus Δu response of such a prism exhibits a negative slope for each examined value of Δu. Differently, the T3R prism shows an always

Table 3.3 Axial stiffness K_0 and coupling parameter $q_0 = \lim_{\Delta u \to 0^+} (\Delta\vartheta/\Delta u)$ in correspondence to the reference configurations of the examined prisms

Prism	K_0 (N/mm)	q_0 (mm/rad)
T3D	12.49	0.40
T3R	13.70	0.43
TθD	57.39	0.40
TθR	61.04	0.43

stiffening F versus Δu response, and a $\Delta\vartheta/\Delta u$ versus Δu curve with a positive slope, which tends to reach a horizontal plateau under large displacements (Fig. 3.1-bottom). Finally, both the analyzed Tθ prisms (TθD and TθR) exhibit an elastically softening elastic F versus Δu response, which is accompanied by a $\Delta\vartheta/\Delta u$ versus Δu curve with always negative slope, in correspondence to the examined range of Δu (Fig. 3.2). The results presented in Fraternali et al. (2012), Fraternali et al. (2014), Fraternali et al. (2015) demonstrate that the aforementioned behaviors significantly impact the wave dynamics of mass-spring systems in the large-displacement regime. For a detailed examination of these dynamics, we refer the reader to the Chapter dedicated to the study of solitary wave dynamics in tensegrity lattices.

The path-following procedures employed to compute the equilibrium paths illustrated in Figs. 3.1 and 3.2 are diffusely illustrated in Fraternali et al. (2015), Mascolo et al. (2018), to which the reader is referred to for the details. For future use, it is worth computing the values of the axial stiffness K_0 and the coupling ratio $q_0 = \lim_{\Delta u \to 0^+} (\Delta\vartheta/\Delta u)$ between twisting rotations and axial displacements, which affect the reference configuration of the prisms. Table 3.3 lists such quantities for the examined T3 and Tθ units.

3.3 Dispersion Relation of Tensegrity Mass-Spring Chains

The present section studies the relation occurring between the angular frequency and the wave number of Bloch waves (Ashcroft and Mermin, 1976) propagating in mass-spring chains of various types, all of which exhibit T3 and/or Tθ prisms alternating with lumped masses. We begin examining *monoatomic chains* obtained by stacking unit cells formed by a single prism and a lumped mass in the longitudinal direction (Sect. 3.3.1). Next, we examine periodic chains formed by unit cells that encompass both T3 and Tθ prisms joined by lumped masses (*diatomic chains*, cf. Sect. 3.3.2).

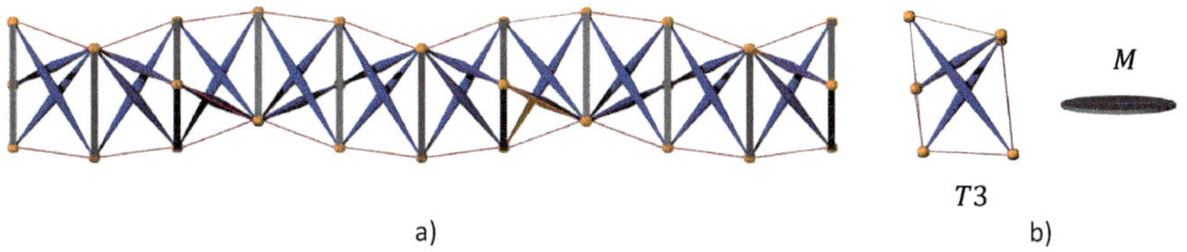

Fig. 3.3 Illustration of a monoatomic mass-spring chain obtained by alternating T3 prisms and lumped masses. **a** Side view of the chain. **b** Close-ups of a lumped mass and a prism unit

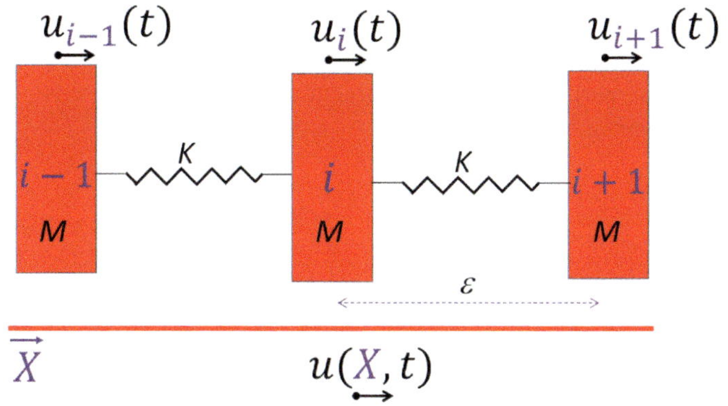

Fig. 3.4 Schematic of a module of a monoatomic chain. Only three masses M and two springs K are shown, i.e. the masses of the unit cells $i-1$, i and $i+1$, and the springs between them. Each mass M is kinematically characterized by the displacement $u_i(t)$ in the axial direction. The generic unit cell is composed of one single mass and one single spring and has edge length (or size) ε. The bottom of the figure shows the continuum one-dimensional domain covered by the chain presented in Sect. 3.4.2. Each point of such a domain has a position X in the reference configuration, and is kinematically characterized by the displacement field $u(X, t)$ in the axial direction

3.3.1 Monoatomic Chains

Monoatomic mass-spring chains with tensegrity architecture are obtained by stacking in sequence T3 (or, alternatively, Tθ) prisms and lumped masses (Fig. 3.3).

We have observed in the previous section that a compression loading of a T3 or a Tθ prism activates relative axial displacements Δu between the terminal bases accompanied by relative twisting rotations $\Delta\vartheta$. One can describe such a coupling deformation process through the parameter $q = (\Delta\vartheta/\Delta u)$. We have indeed introduced the reference value of such a coupling parameter in Table 3.3 of the previous section. We now wish to study the dynamics of a tensegrity mass-spring chain through the model illustrated in Fig. 3.4, which consists of a sequence of $N+1$ masses M connected with linear springs featuring the same stiffness constant K. Such a model is intended to describe the response of a tensegrity chain under a dynamical loading condition that induces small perturbations of a given equilibrium configuration. In particular, we will identify the initial placement of the chain with that formed by the reference configurations of the tensegrity prisms studied in the previous section. Accordingly, we will set the spring constant K equal to the tangent stiffness K_0 exhibited by the prism in such a configuration, and will characterize the coupling between axial and twisting motions through the parameter $q = q_0$.

With reference to the model of Fig. 3.4, we let ε denote the distance between the centers of two adjacent masses, which characterizes the size of a single unit cell. In the reference configuration, the position X_i of the generic mass is given by $X_i = i\varepsilon$, while the deformed position of such an element is given by $x_i = X_i + u_i$, where the displacement u_i is a function of time t. Let us assume that the generic mass consists of a disc with radius r and thickness d, which is made of a material with volumetric mass density ρ_v. Such an assumption leads us to compute the mass M and the rotational inertia J of such an element with respect to the axis of the chain through the following formulae

$$M = \pi r^2 d \rho, \qquad J = \frac{1}{2} M r^2. \tag{3.1}$$

By accounting for the coupling between the longitudinal and the twisting motions, we compute the kinetic energy $\mathcal{K}_i$ of the i-th cell though the following equation

$$\mathcal{K}_i = \frac{1}{2} M \dot{u}_i^2 + \frac{1}{2} J \dot{\theta}_i^2 \tag{3.2}$$

where $\dot{u}_i$ and $\dot{\vartheta}_i$ are the axial and rotational velocities of the i-the mass, which are given by the derivatives with respect to time of the functions u_i and of ϑ_i, respectively. Hereafter we convene to employ the dot notation for time derivatives, by letting $\dot{f}$ denote the first-order derivative of the function of time f, and $\ddot{f}$ the second-order time derivative of the same function. The linearization of the mechanical responses of the prisms forming the chain, from the reference configuration, allows us to write

$$\vartheta_{i+1} - \vartheta_i = q\,(u_i - u_{i+1}), \qquad \forall i = 0, ..., N-1 \tag{3.3}$$

If now we assume $u_N = 0$ and $\vartheta_N = 0$ (vanishing boundary conditions at the end of the chain, for both the axial displacement and the twisting rotation), it is easy to show that Eq. (3.3) leads us to the result

$$\vartheta_i = -q\, u_i, \qquad \forall i = 0, ..., N \tag{3.4}$$

which implies that we can fully characterize the kinematics of the unit cell through its longitudinal displacement u_i. We can therefore cast Eq. (3.2) into the following form

$$\mathcal{K}_i = \frac{1}{2} M \dot{u}_i^2 + \frac{1}{2} J q^2 \dot{u}_i^2 = \frac{1}{2} M^{eq} \dot{u}_i^2 \tag{3.5}$$

where we have introduced the *equivalent mass*

$$M^{eq} = M\left(1 + Jq^2\right) = M\left(1 + \frac{q^2 r^2}{2}\right) \tag{3.6}$$

Let us now denote the overall kinetic, internal and external energies of the chain by $\mathcal{K}$, $\mathcal{U}_{int}$ and $\mathcal{U}_{ext}$, respectively. Such energies are obtained by summing up the contributions of all the unit cells, that is

$$\mathcal{K} = \sum_{i=0}^{N} \frac{1}{2} M^{eq} \dot{u}_i^2, \qquad \mathcal{U}_{int} = \sum_{i=0}^{N} \frac{1}{2} K \left(u_{i+1} - u_i\right)^2, \qquad \mathcal{U}_{ext} = \sum_{i=0}^{N} \mathcal{F}_i^{ext,eq} u_i, \tag{3.7}$$

where $\mathcal{F}_i^{ext,eq}$ is the *equivalent external force* acting on the i-th mass, which is given by

$$\mathcal{F}_i^{ext,eq} = F_i^{ext} + \mathcal{M}_i^{ext} q \tag{3.8}$$

F_i^{ext} and $\mathcal{M}_i^{ext}$ denoting the external force and the external torque acting on the i-th mass, respectively.

The *Action functional* of the system is defined in the following standard way (see, e.g., Landau and Lifshitz 1976)

$$\mathcal{A} = \int_{t_1}^{t_2} \left[\mathcal{K} - \mathcal{U}_{int} + \mathcal{U}_{ext}\right] dt, \tag{3.9}$$

where t_1 and t_2 are two instants of time for which the displacements u_i of all the masses are prescribed, that is

$$u_i(t = t_1) = u_1, \qquad u_i(t = t_2) = u_2, \qquad \forall i = 0, ..., N. \tag{3.10}$$

A variational principle

$$\delta\mathcal{A} = 0, \qquad \forall \delta u_i \in \mathcal{C}_i \qquad \forall i = 0, ..., N, \tag{3.11}$$

is assumed to be valid for any admissible variations $\mathcal{C}_i$ of u_i (Landau and Lifshitz, 1976). The equations of motion of the chain are derived from the action principle (3.11) once the restrictions (3.10) are enforced, obtaining

$$M^{eq} \ddot{u}_i = \mathcal{F}_i^{ext,eq} + K(u_{i+1} - 2u_i + u_{i-1}) \qquad \forall i = 0, ..., N, \qquad \forall t \in \mathbb{R}. \tag{3.12}$$

We now follow the Bloch-Floquet theory of discrete systems, by looking for propagating harmonic wave solutions of the following form (Ashcroft and Mermin, 1976)

$$u_i = \mathrm{Re}\left\{u_0 \exp\left[I\left(\omega t - k_w i \varepsilon\right)\right]\right\}, \tag{3.13}$$

under a small dynamic perturbation of the reference configuration of the chain. In Eq. (3.13), ω is the angular frequency of the traveling wave, k_w is the wave number (i.e., 2π times the inverse of the wavelength λ), I is the imaginary unit, Re is the real part operator and u_0 is the wave amplitude. In the absence of external forces, which implies

$$\mathfrak{F}_i^{ext,eq} = 0, \qquad \forall i = 0, ..., N, \tag{3.14}$$

the insertion of (3.13) and (3.14) into the ordinary differential equations (3.12) yields the following dispersion relation between the angular frequency ω and the wave number k_w of Bloch wavefunctions traveling through the chain

$$\omega = \omega_m(k_w) = \sqrt{\frac{K}{M^{eq}}}\sqrt{2(1 - \cos k_w \varepsilon)} = 2\sqrt{\frac{K}{M^{eq}}}\left|\sin \frac{k_w \varepsilon}{2}\right| \tag{3.15}$$

The subscript m appearing in Eq. (3.15) stands for monoatomic. A physical example of a monoatomic chain is offered by a chain of T3R prisms alternated with lumped mass disks, which are made of lead ($\rho_v = 11340$ kg/m^3) and exhibit a diameter of 15 mm and 2 mm thickness. The mass M of such disks amounts to 4 g. We use T3R prisms since we are assuming that the bases of the prisms are perfectly bonded to the lumped masses, which implies that such members transform rigidly during the motion of the chain (see also the observations made in the previous section). Assuming $K =$ 13.70 N/mm and $q = 0.43$ (cf. Table 3.3), we obtain $M^{eq} = 24.85$ g from Eq. (3.6). The dispersion relation (3.15) of the present monoatomic chain is pictured in Fig. 3.5, where the linear frequency $\nu = \omega/(2\pi)$ of the traveling wave has been introduced. Since the solution given by Eq. (3.15) is periodic with period $2\pi/\varepsilon$, in terms of the variable k_w, the plot of Fig. 3.5 has been restricted to the *first Brillouin zone* (Ashcroft and Mermin, 1976), which is such that $k_w\varepsilon$ varies between 0 and $2\pi \approx 6.283$.

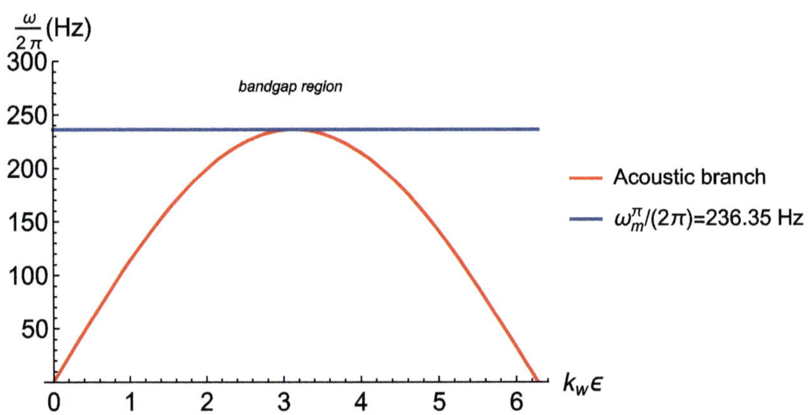

Fig. 3.5 Dispersion relation of a monoatomic chain of T3 prisms that exhibits coupled longitudinal and twisting motions

It is worth to note that the wave dynamics of the the monoatomic chain is characterized by one degree of freedom per cell, and that the dispersion relation exhibits only one 'acoustic' branch ω_m defined by Eq. (3.15). It is easily seen that no signal of the form (3.13) can propagate through the chain at a frequency higher than the frequency ω_m^π, i.e.

$$\omega > \omega_m^\pi = 2\,\omega_m^r = 2\sqrt{\frac{K}{M^{eq}}}, \qquad \Rightarrow \qquad \textit{no wave solution}, \tag{3.16}$$

where ω_m^r is the resonant frequency associated to the single oscillator forming the generic unit cell, with stiffness K and mass M^{eq}. The frequency range with $\omega > \omega_m^\pi$ is referred to as the *bandgap region* of the monoatomic chain.

Let us now define the *phase velocity* v_p and the *group velocity* v_g of the waves traveling through the chain, which are respectively given by (Ashcroft and Mermin, 1976)

$$v_p = \frac{\omega}{k_w}, \qquad v_g = \frac{d\omega}{dk_w} \tag{3.17}$$

Such quantities represent the speed of a single traveling wave, and the speed of packets of waves propagating across the chain, respectively. In particular, the group velocity represents the velocity at which the mechanical energy propagates through the chain (Ashcroft and Mermin, 1976). Using Eq. (3.15), we obtain for the chain under consideration

$$v_g = \varepsilon\sqrt{\frac{K}{M^{eq}}}\cos\frac{k_w\,\varepsilon}{2} \tag{3.18}$$

One notes that it results in $v_g = 0$ for $k_w\varepsilon = \pi$, which implies that at the edge of the bandgap region the velocity of transmission of mechanical energy is zero. Since Eq. (3.15) implies that $d\omega/dk_w$ is a non-constant function of ω, the present chain is said to be *dispersive*. Such a nomenclature derives from the observation that the different components of a wave packet (formed by waves traveling with different frequencies) move along the chain with different velocities. This causes the spread over time (i.e., the *dispersion*) of the wave packet. Section 3.4.2 will study what happens to the dispersion relation (3.15) in the *long wavelength limit* corresponding to the propagation of waves such that the wave number k_w is much smaller than $1/\varepsilon$ (i.e, the wavelength is much larger than the spacing ε of the masses forming the chain).

The frequency bandgap of the T3 chain under examination corresponds to the linear frequency range $\omega/(2\pi) > \omega_m^\pi/(2\pi) = 236.35$ Hz, as it is shown in Fig. 3.5. A different response is obtained when one considers the same chain but makes the assumption that now the T3 prisms are in frictionless contact with the lumped masses, in such a way that the masses do not exhibit rotations about the axis of the chain, but only axial displacements (uncoupled chain: see, e.g., Amendola et al. 2018). This situation can be regarded as a special case of the study conducted so far, which is obtained by setting $M^{eq} = M = 4$g in the dispersion relation (3.15). As a result one obtains $\omega_m^\pi/(2\pi) = 588.52$ Hz, that is a lower bound of the bandgap region much higher than that corresponding to the coupled chain. A slightly different response is obtained for a new uncoupled chain that is equipped with T3D prisms instead of T3R prisms. Assuming $K = 12.49$ N/mm (cf. Table 3.3) and $M^{eq} = 4$ g we obtain $\omega_m^\pi/(2\pi) = 562.02$ Hz for this other chain. It can be observed that one can actually predict a large variety of bandgap responses of the tensegrity monoatomic chain under examination, by varying the state of internal and external prestress of the system (i.e., the initial configuration), in addition to the modification of the geometry, topology and sizing of bars and cables (Amendola et al., 2018).

3.3.2 Diatomic Chains

Figure 3.6 shows a diatomic mass-spring chain formed by alternating T3 prisms (hereafter referred to as P_1 units), Tθ prisms (P_2 units) and lumped (disk) masses. Under small compression perturbations of the reference configuration, we describe the dynamic response of such a system through the mechanical model illustrated in Fig. 3.7.

The current chain is a sequence of $2N$ masses M_1 and M_2 connected with linear springs featuring alternating stiffness constants K_1 and K_2 (Fig. 3.7). The latter are let to coincide with the tangent axial stiffness coefficients of the prism units P_1 and P_2 in correspondence to the reference configuration of the chain. The overall edge length ε of the unit cell is equal to the summation of the equilibrium length ε_1 of the P_1 prism and the equilibrium length ε_2 of the P_2 prism. Examining the i-th unit cell, we observe that the reference position of the mass M_1 is given by $X_i^1 = i\varepsilon + \varepsilon_1$, while the reference position of the mass M_2 is given by $X_i^2 = i\varepsilon$. The current (deformed) position of the same masses are instead given by $x_i^1 = X_i^1 + u_i^1$ and $x_i^2 = X_i^2 + u_i^2$, respectively, where u_i^1 and u_i^2 are functions of time t. We pause to observe that we

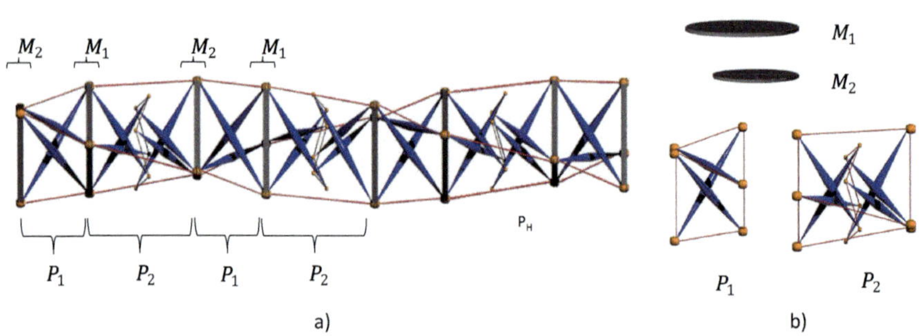

Fig. 3.6 Illustration of a diatomic mass-spring chain with tensegrity architecture, which is obtained by alternating T3 prisms, Tθ prisms and lumped masses. **a** Side view of the chain. **b** Close-ups of the lumped masses and the prism units

are using a superscript notation for the displacements of the masses M_1 and M_2 forming the generic unit cell. As a consequence, in order to avoid misinterpretations of the following mathematical passages, we will often adopt the bracket notation for exponentiation of such displacement variables, by letting, e.g., $(u_i^2)^n$ indicate the displacement u_i^2 raised to the power of n.

Let us denote the rotations about the chain axis of the masses M_1 and M_2 of the i-th cell by ϑ_i^1 and ϑ_i^2, respectively. The kinetic energy of such a cell is computed as follows

$$\mathcal{K}_i = \frac{1}{2}\left[M_1(\dot{u}_i^1)^2 + J_1(\dot{\vartheta}_i^1)^2 + M_2(\dot{u}_i^2)^2 + J_2(\dot{\vartheta}_i^2)^2\right] \tag{3.19}$$

where we have again used the dot notation for time derivatives. By linearizing the mechanical response of the prisms P_1 and P_2 in the proximity of the reference configuration, we write

$$\vartheta_i^2 - \vartheta_{i-1}^1 = q_2\,(u_{i-1}^1 - u_i^2),\quad \vartheta_i^1 - \vartheta_i^2 = q_1\,(u_i^2 - u_i^1),\quad \forall i = 0, \ldots, N-1 \tag{3.20}$$

having denoted the coupling parameters of the rotational and twisting motions of the prisms P_1 and P_2 by q_1 and q_2, respectively. We now assume vanishing boundary

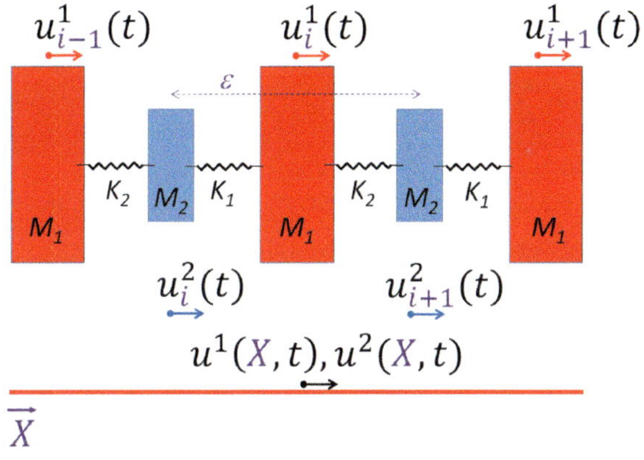

Fig. 3.7 Schematic of a module of a diatomic chain. Only five masses (three masses M_1 and two masses M_2) and four springs (two with stiffness constant K_1 and two with stiffness constant K_2) are shown. The present sketch indeed includes the two masses of the unit cells i and $i+1$, and the mass M_1 of the unit cell $i-1$. The M_1 masses are kinematically characterized by the displacements $u_i^1(t)$ in the axial direction, while the masses M_2 are characterized by the axial displacements $u_i^2(t)$. Overall, the generic unit cell is composed of two masses and two springs and has a total size ε. The bottom of the figure illustrates the continuum one-dimensional domain covered by the chain presented in Sect. 3.4.3. The generic point point of such a domain has a position X in the reference configuration and is kinematically characterized by two displacement fields $u^1(X,t)$ and $u^2(X,t)$ in the axial direction

conditions at the end of the chain for the axial displacement u_N^2 and the rotation ϑ_N^2 of the mass M_2. In addition, we further assume that it results $q = q_1 = q_2$. The adoption of such simplifying hypotheses allows us to derive from Eq. (3.20) the following results

$$\vartheta_i^1 = -q\, u_i^1, \qquad \vartheta_i^2 = -q\, u_i^2, \qquad \forall i = 0, ..., N-1 \tag{3.21}$$

Going back to Eq. (3.19), we can now write

$$\mathcal{K}_i = \frac{1}{2}\left[M_1^{eq}(\dot{u}_i^1)^2 + M_2^{eq}(\dot{u}_i^2)^2\right] \tag{3.22}$$

where we have introduced the following equivalent masses

$$M_1^{eq} = M_1\left(1 + J_1 q^2\right), \qquad M_2^{eq} = M_2\left(1 + J_2 q^2\right) \tag{3.23}$$

The overall kinetic, internal and external energies of the diatomic chain are computed through

$$\mathcal{K} = \sum_{i=0}^{N-1} \frac{1}{2}\left[M_1^{eq}(\dot{u}_i^1)^2 + M_2^{eq}(\dot{u}_i^2)^2\right], \tag{3.24}$$

$$\mathcal{U}_{int} = \sum_{i=0}^{N-1} \frac{1}{2}\left[K_1\left(u_i^1 - u_i^2\right)^2 + K_2\left(u_{i+1}^2 - u_i^1\right)^2\right], \tag{3.25}$$

$$\mathcal{U}_{ext} = \sum_{i=0}^{N-1}\left[\mathcal{F}_{1,i}^{ext,eq} u_i^1 + \mathcal{F}_{2,i}^{ext,eq} u_i^2\right], \tag{3.26}$$

where the following equivalent external forces have been introduced

$$\mathcal{F}_{i,1}^{ext,eq} = F_{i,1}^{ext} + \mathcal{M}_{i,1}^{ext} q, \qquad \mathcal{F}_{i,2}^{ext,eq} = F_{i,2}^{ext} + \mathcal{M}_{i,2}^{ext} q \tag{3.27}$$

$F_{1,i}^{ext}$, $F_{2,i}^{ext}$, $\mathcal{M}_{i,1}^{ext}$ and $\mathcal{M}_{i,2}^{ext}$ denoting the external forces and moments acting on the masses forming the i-th unit cell. The Action functional of the diatomic system is defined in the standard way (3.9), where t_1 and t_2 are now two instants of time for which the displacements u_i^1 and u_i^2 of all the masses are prescribed, which results in

$$u_i^1(t = t_A) = u_A^1, \quad u_i^2(t = t_A) = u_A^2, \quad A = 1, 2, \quad \forall i = 0, ..., N \tag{3.28}$$

Also in the case of the diatomic chain, as well as in the case of the monoatomic chain analyzed in the previous section, a variational principle of the form

$$\delta\mathcal{A} = 0, \qquad \forall \delta u_i^1 \in \mathcal{C}_i^1, \quad \forall \delta u_i^2 \in \mathcal{C}_i^2 \qquad \forall i = 0, ..., N \tag{3.29}$$

is assumed to be valid for any admissible variations $\mathcal{C}_i^1$ of u_i^1 and $\mathcal{C}_i^2$ of u_i^2. The equations of motion of the chain are derived from (3.29) once the restrictions (3.28) are enforced, which leads us to

$$M_1^{eq}\ddot{u}_i^1 = \mathcal{F}_{1,i}^{ext,eq} - (K_1 + K_2)u_i^1 + K_1 u_i^2 + K_2 u_{i+1}^2, \ \forall i = 0, ..., N-1, \ \forall t \in \mathbb{R} \quad (3.30)$$
$$M_2^{eq}\ddot{u}_i^2 = \mathcal{F}_{2,i}^{ext,eq} - (K_1 + K_2)u_i^2 + K_1 u_i^1 + K_2 u_{i-1}^1, \ \forall i = 1, ..., N, \ \forall t \in \mathbb{R} \quad (3.31)$$

We now search for Bloch-wave solutions of the following form

$$u_i^1 = \text{Re}\left\{u_0^1 \exp\left[I\left(\omega t - k_w i\varepsilon\right)\right]\right\}, \quad u_i^2 = \text{Re}\left\{u_0^2 \exp\left[I\left(\omega t - k_w i\varepsilon\right)\right]\right\} \quad (3.32)$$

where u_0^1 and u_0^2 are the two characteristic wave amplitudes of the diatomic chain. In the absence of external forces ($\mathcal{F}_{1,i}^{ext,eq} = \mathcal{F}_{2,i}^{ext,eq} = 0, \ \forall i = 0, ..., N$), the insertion of Eq. (3.32) into the ordinary differential equations (3.30)–(3.31) yields the following dispersion relation

$$\omega_{do,da}\left(k_w\right) = \sqrt{\frac{(K_1 + K_2)(M_1^{eq} + M_2^{eq}) \pm \Delta_d\left(k_w\right)}{2M_1^{eq}M_2^{eq}}} \quad (3.33)$$

where we have set

$$\Delta_d\left(k_w\right) = \sqrt{\left[(K_1 + K_2)(M_1^{eq} + M_2^{eq})\right]^2 - 8M_1^{eq}M_2^{eq}K_1K_2(1 - \cos k_w\varepsilon)} \quad (3.34)$$

The subscripts *do* and *da* in (3.33) stand for optic and acoustic branches of the diatomic dispersion relation, respectively.

Figure 3.8 graphically illustrates the dispersion relations (3.33) of two different diatomic chains. Both such chains show T3R (P_1) prisms alternated with Tθ (P_2) prisms, which implies $K_1 = 13.70$ N/mm and $K_2 = 61.04$ N/mm (cf. Table 3.3). In the chain #1 we use the same mass disks employed for the monoatomic chains of the previous section (lead disks with 15 mm diameter and 2 mm thickness), for both the mass M_1 and the mass M_2 ($M_1^{eq} = M_2^{eq} = 24.85$ g). This chain therefore shows only stiffness contrast between the spring elements of the unit cell, and no mass contrast. The chain #2 instead features $M_1^{eq} = 24.85$ g and $M_2^{eq} = 73$ g, due to the use of lead disks with 20 mm diameter and 2 mm thickness to form the M_2 masses. In both cases, we assumed $q = q_1 = q_2 = 0.43$ (Table 3.3).

It is worth to note that in the diatomic chain the dynamics of the system is characterized by two degrees of freedom per unit cell, and therefore the dispersion relation is formed by two different branches, one named 'acoustic' ω_{da} (lower branch) and the other one named 'optic' ω_{do} (upper branch). For $k_w = 0$, it is easily shown that the optic branch assumes the following nonzero value

$$\omega_{do}\left(0\right) = \omega_{do}^0 = \sqrt{(K_1 + K_2)\frac{M_1^{eq} + M_2^{eq}}{M_1^{eq}M_2^{eq}}}. \quad (3.35)$$

It is also not difficult to verify that the acoustic branch tends to assume the following linear expression for small values of k_w

$$\omega_{da}(k_w) \cong k_w \sqrt{\frac{K_1 K_2}{(K_1 + K_2)(M_1^{eq} + M_2^{eq})}} \tag{3.36}$$

and one has $\omega_{Ma} = 0$ for $k_w = 0$. Differently, for $k_w = \pi/\varepsilon$, both the acoustic branch and the optic branch assume nonzero values, which we label $\omega_{da}(\pi/\varepsilon) = \omega_{da}^{\pi}$ and $\omega_{do}(\pi/\varepsilon) = \omega_{do}^{\pi}$, respectively. It is seen from the examples illustrated in Fig. 3.8 that the dispersion relation (3.33) opens two bandgap regions, since it results

$$\omega \in \left(\omega_{da}^{\pi}, \omega_{do}^{\pi}\right) \, U \, \left(\omega_{do}^{0}, +\infty\right), \quad \omega_{do}^{\pi} \leq \omega_{do}^{0}, \quad \Rightarrow \quad \textit{no wave solution} \tag{3.37}$$

The upper bandgap region ($\omega \in \left(\omega_{do}^{0}, +\infty\right)$) is similar to that of the monoatomic case (3.16). It prescribes no Bloch-waveform solutions for frequencies greater than

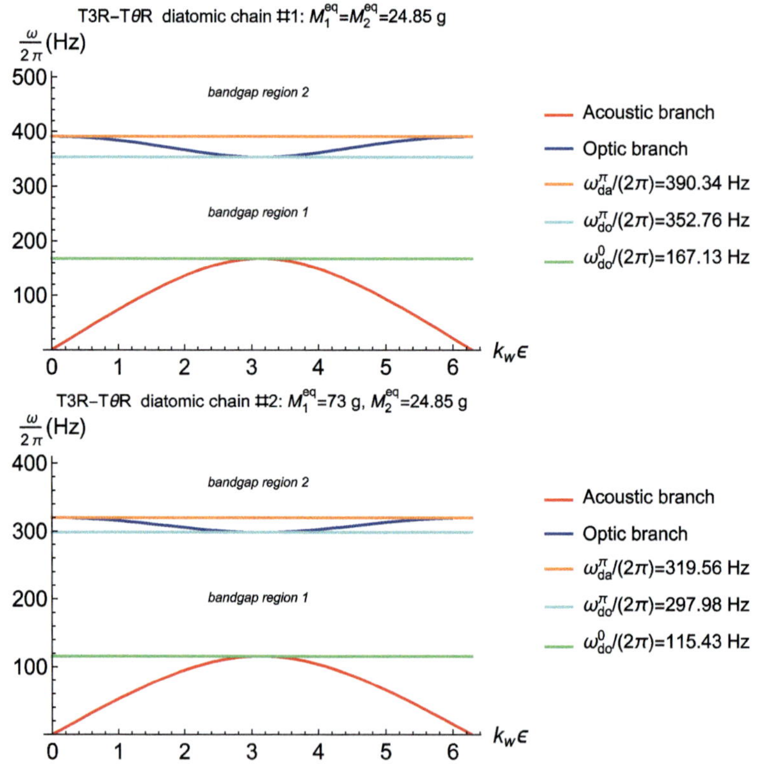

Fig. 3.8 Dispersion relations of two diatomic chains featuring T3R prisms alternated with TθR prisms

the resonance frequency of the two-body system, elastically connected with the parallel $K_1 + K_2$ of the two stiffness, with reduced mass $M_1^{eq} M_2^{eq}/(M_1^{eq} + M_2^{eq})$. Besides, the lower bandgap region ($\omega \in \left(\omega_{da}^{\pi}, \omega_{do}^{\pi}\right)$) in (3.37) is often referred to as the primary bandgap of the diatomic chain. Such a bandgap originates by the mass contrast and/or the stiffness contrast between the elements of the unit cell. It disappears for $M_1^{eq} = M_2^{eq}$ and $K_1 = K_2$, when the dispersion relation of the diatomic chain becomes coincident with that of a monoatomic chain.

3.4 Continuum Models

3.4.1 Motivation for Continuum Approaches

In the case of an externally forced system, when the length of the chain is finite and/or when the presence of inhomogeneities breaks the periodicity of the system, the waveform solutions (3.13) and (3.32) do not longer satisfy the equations of motion. Such an inconvenience can be tackled mainly through the following two strategies. The first one consists of solving the N (or the $2N$) ordinary differential equations (3.12) (or (3.30)–(3.31)) with the competent initial and boundary conditions, via numerical integration techniques (refer, e.g., to Fraternali et al. 2012; Fraternali et al. 2014). Such a strategy is generally accurate from the numerical point of view, but has a computational disadvantage if the number of unit cells (or 'atoms') forming the chain is very large. A second available strategy makes recourse to the derivation of a continuum model of the chain, equipped with suitable Partial Differential Equations (PDEs) and Boundary Conditions (BCs). Such an approach is accurate only when the wavelength $\lambda = 2\pi/k_w$ is sufficiently larger than the size ε of the unit cell. It offers, however, the advantage to require a computational effort that is either negligible, when analytic solutions are available, or affordable in the case, e.g., of finite element approximations.

The present section develops continuum models for the monoatomic and diatomic systems illustrated in the previous section. We adopt a discrete-to-continuum approach based on the introduction of suitable, continuous displacement fields (see the bottom portions of Figs. 3.4 and 3.7), and the assumption of the following Piola's ansatz (Barchiesi and Khakalo, 2019; Turco et al., 2018; Barchiesi et al., 2019; Turco et al., 2020)

$$f(X_i) = f_i, \qquad \forall i = 0, ..., N, \quad \varepsilon \ll L \quad \Rightarrow \quad \sum_{i=0}^{N} f_i \varepsilon = \int_0^L f(X) dX, \tag{3.38}$$

Here, f_i is a quantity defined on a given unit cell i of the discrete model, which Eq. $(3.38)_1$ relates to a continuous field $f(X)$ defined on a certain domain ($X \in [0, L]$), in such a way that the Riemann sum equation $(3.38)_2$ holds, for $\varepsilon \ll L$.

3.4.2 A Continuum Model of the Monoatomic Chain

Let us introduce a one-dimensional continuum domain of length L equipped with a displacement field $u(X,t)$ acting as a kinematic descriptor (see Fig. 3.4), where $X \in [0, L]$ and $t \in [t_1, t_2]$. We use a Piola's ansatz of the form (3.38) to relate the displacements of the unit cells of the discrete model to such a field, that is

$$u(X_i, t) = u_i(t), \qquad \forall i = 0, ..., N, \qquad \forall t \in [t_1, t_2] \tag{3.39}$$

Equation (3.39) relates the displacement $u_i(t)$ of the generic mass forming the discrete model to the value taken by the continuous field descriptor $u(X,t)$ at the point $X = X_i$, for every time $t \in [t_1, t_2]$, and for every unit cell $i = 0, ..., N$. The adoption of Eqs. (3.39) and (3.38) (with $f = u$ and $f_i = u_i$) leads us to write the kinetic energy of the monoatomic chain into the following form

$$\mathcal{K} = \sum_{i=1}^{N} \frac{1}{2} M^{eq} \left(\dot{u}_i(t)\right)^2 \frac{\varepsilon}{\varepsilon} = \sum_{i=1}^{N} \frac{1}{2} \frac{M^{eq}}{\varepsilon} \left(\dot{u}(X_i, t)\right)^2 \varepsilon = \int_0^L \frac{1}{2} \rho \dot{u}^2 dX, \tag{3.40}$$

where the linear mass density ρ of the continuous model has been introduced through

$$\rho = \frac{M^{eq}}{\varepsilon} \tag{3.41}$$

M^{eq} denoting the equivalent mass of the discrete model. The latter accounts for the coupling between twisting and axial motions, as we have seen in the previous section. In the limit for $\varepsilon \to 0$, Eq. (3.41) introduces a well-defined identification of ρ, assuming that M^{eq} rescales with the length ε of the unit cell, when one considers a sequence of discrete models that is parametrized by values of ε tending to zero. In other words, we are assuming that the equivalent mass M^{eq} scales with ε in the limit $\varepsilon \to 0$, so that the linear mass density ρ retains a finite value. It is worth noting that the adoption of Eq. (3.41) will lead us to the development of a continuous model of a beam undergoing axial deformations, which implicitly accounts for the coupling between twisting and longitudinal deformations of the monoatomic chain.

Moving on to define the internal energy $\mathcal{U}_{int}$ of such a continuous model, we now introduce the following Taylor's series expansion of $u(X_{i+1}, t)$ centered at $X = X_i$

$$u_{i+1}(t) = u(X_{i+1}, t) = u(X_i + \varepsilon, t) \cong u(X_i, t) + \varepsilon u'(X_i, t) + \frac{1}{2} \varepsilon^2 u''(X_i, t) \tag{3.42}$$

Making use of Eqs. (3.42), (3.39), and $(3.38)_2$ we write

$$\mathcal{U}_{int} = \sum_{i=1}^{N} \frac{1}{2} K \left(u_{i+1} - u_i\right)^2 \frac{\varepsilon}{\varepsilon} = \sum_{i=1}^{N} \frac{1}{2} \frac{K}{\varepsilon} \left(\varepsilon u'(X_i, t) + \frac{1}{2} \varepsilon^2 u''(X_i, t)\right)^2 \varepsilon$$

$$= \int_0^L \frac{1}{2} K_N \left(u'(X_i, t) + \frac{\varepsilon}{2} u''(X, t) \right)^2 dX \tag{3.43}$$

where the axial stiffness K_N of the continuous model is defined as follows

$$K_N = K\,\varepsilon \tag{3.44}$$

in terms of the stiffness coefficient K of the discrete model. Equation (3.44) introduces a well-defined identification of K_N provided that K scales with the reciprocal of ε, which implies $K \to \infty$ for $\varepsilon \to 0$, in such a way that the axial stiffness K_N retains a finite value. In the same limit, we deduce from (3.43) that the second gradient term $\varepsilon u''$ on the right hand side tends to zero, so that the internal strain energy reduces to that of a standard beam with axial stiffness (3.44). However, for finite (but small) values of the length ε of the unit cell, a second gradient elastic energy of the form (3.43) can be usefully considered.

A second gradient internal energy calls for the introduction of external actions that include the dual of the displacement gradient (i.e., an external 'double force'). In addition, the external actions of the continuous model must include both distributed and concentrated forces. On accounting for such requirements, we formulate the following expression of the external energy of the continuous model

$$\mathcal{U}_{ext} = \int_0^L \left(b_N u + d_N u' \right) dX + F_0 u(0, t) + F_L u(L, t) + D_0 u'(0, t) + D_L u'(L, t) \tag{3.45}$$

where we have introduced the external distributed force $b_N(X, t)$; the external distributed double force $d_N(X, t)$; the external concentrated forces at the extremities (or bases) of the continuous medium $F_0(t)$, $F_L(t)$, and the external concentrated double forces at such extremities $D_0(t)$ and $D_L(t)$. The forces on the bases are defined through the following identification formulae

$$\mathcal{F}_0^{ext,eq} = F_0, \quad \mathcal{F}_N^{ext,eq} = F_L, \quad \mathcal{F}_1^{ext,eq} - \mathcal{F}_0^{ext,eq} = \varepsilon D_0, \quad \mathcal{F}_N^{ext,eq} - \mathcal{F}_{N-1}^{ext,eq} = \varepsilon D_L, \tag{3.46}$$

where first-order derivatives have been approximated through finite differences. Similarly, the distributed external actions are defined through

$$\mathcal{F}_i^{ext,eq} = \varepsilon b_N(X_i, t), \quad \mathcal{F}_{i+1}^{ext,eq} - \mathcal{F}_i^{ext,eq} = \varepsilon^2 d_N(X_i, t), \quad \forall i = 0, ..., N-1, \tag{3.47}$$

having used a discrete approximation of the integral appearing in Eq. (3.45). The Action functional of the continuous system is always defined in the form (3.9), and the equations of motion of such a system are derived from the action principle

$$\delta \mathcal{A} = 0, \qquad \forall \delta u \in \mathcal{C}, \tag{3.48}$$

for any admissible variations $\mathcal{C}$ of u. The Piola's ansatz (3.39) and the boundary conditions (3.10) of the discrete model require

$$u(X, t_1) = u_1(X), \quad u(X, t_2) = u_2(X) \tag{3.49}$$

which enforce restrictions also on the boundaries of the time domain. We derive the PDEs and BCs that rule the continuum model from the action principle (3.48), by computing the variation of the action functional as follows

$$\delta \mathcal{A} = \int_{t_1}^{t_2} \left[\int_0^L \left[\delta u \left(K_N \frac{\varepsilon^2}{4} u'''' - K_N u'' - b_N + d_N' + \rho \ddot{u} \right) \right] dX \right] dt \tag{3.50}$$

$$+ \left[\delta u \left(K_N u' - \frac{\varepsilon^2}{4} u''' - d_N \right) \right]_0^L + \left[\delta u' \left(\frac{\varepsilon}{2} u' + \frac{\varepsilon^2}{4} u'' \right) \right]_0^L \tag{3.51}$$

$$- F_0 \delta u(0, t) - F_L \delta u(L, t) - D_0 \delta u'(0, t) - D_L \delta u'(L, t) = 0. \tag{3.52}$$

The arbitrariness of the variations appearing in Eq. (3.48) leads us to the the following PDE

$$K_N \frac{\varepsilon^2}{4} u'''' - K_N u'' - b_N + d_N' + \rho \ddot{u} = 0, \qquad \forall X \in [0, L], \tag{3.53}$$

to be satisfied under the following BCs

$$u(0, t) = u_0(t) \quad or \quad K_N u'(0, t) - \frac{\varepsilon^2}{4} u'''(0, t) - d_N(0, t) = -F_0(t) \tag{3.54}$$

$$u(L, t) = u_L(t) \quad or \quad K_N u'(L, t) - \frac{\varepsilon^2}{4} u'''(L, t) - d_N(L, t) = F_L(t) \tag{3.55}$$

$$u'(0, t) = u_0'(t) \quad or \quad \frac{\varepsilon}{2} u'(0, t) + \frac{\varepsilon^2}{4} u''(0, t) = -D_0 \tag{3.56}$$

$$u'(L, t) = u_L'(t) \quad or \quad \frac{\varepsilon}{2} u'(L, t) + \frac{\varepsilon^2}{4} u''(L, t) = D_L \tag{3.57}$$

Here, we have listed kinematic BCs on the left and static Bcs on the right. By assuming now the absence of external actions (i.e., $b_N = 0$ and $d_N = 0$), and a plane wave solution for the displacement field $u(X, t)$ of the form

$$u = \text{Re} \{ u_0 \exp [I (\omega t - k_w X)] \}, \tag{3.58}$$

where u_0, ω and k_w are, respectively, the complex amplitude, the angular frequency and the wave number, we obtain the following dispersion relation for the monoatomic continuous medium ('Mc')

$$\omega = \omega_{Mc} (k_w) = k_w \sqrt{\frac{K_N}{\rho} \left(1 + \frac{\varepsilon^2}{4} k_w^2 \right)}, \tag{3.59}$$

Equation (3.59) defines a nonlinear relationship describing a dispersive continuum medium (cf. the observations made at the end of Sect. 3.3.1). However, in limit $\varepsilon \to 0$, such an equation reduces to the following linear form

$$\omega = \omega_{Mc}(k_w) \cong k_w \sqrt{\frac{K_N}{\rho}}. \tag{3.60}$$

that describes the non dispersive relation of a standard beam undergoing axial deformations (Ashcroft and Mermin, 1976). As a result, the velocity of the propagating waves does not depend on the frequency ω in such a medium, being an intrinsic mechanical property that is usually referred to as the *speed of sound* $c = \omega / k_w$. Using Eqs. (3.60), (3.41) and (3.44), we easily obtain $c = \sqrt{K_N/\rho} = \varepsilon\sqrt{K/M^{eq}}$.

3.4.3 A Continuum Model of the Diatomic Chain

We move on to develop a continuum model of the diatomic chain, by introducing a continuous one-dimensional domain of length L, which is characterized by two displacement fields $u_1(X, t)$ and $u_2(X, t)$ (see Fig. 3.7), where $X \in [0, L]$ and $t \in [t_1, t_2]$. The diatomic model actually consists of a mixture of two solids. The first solid describes the response of the masses M_1^{eq}, while the second one describes the response of the masses M_2^{eq}. Both such solids are defined not only for every time $t \in [t_1, t_2]$, but also for every point $X \in [0, L]$ of the reference configuration. Each of them exhibits its own dynamics: one being defined by the displacement field $u_1(X, t)$, and the other one being defined by the displacement field $u_2(X, t)$. The development of the diatomic continuum model originates by the Piola's ansatz

$$u_1(X_i, t) = u_i^1(t), \qquad u_2(X_i, t) = u_i^2(t), \qquad \forall i = 0, ..., N. \tag{3.61}$$

Using Eqs. (3.38) and (3.61), we write the kinetic energy of the current model as follows

$$\mathcal{K} = \sum_{i=0}^{N} \varepsilon \frac{1}{2} \left[\frac{M_1^{eq}}{\varepsilon} (\dot{u}_i^1)^2 + \frac{M_2^{eq}}{\varepsilon} (\dot{u}_i^2)^2 \right] = \int_0^L \frac{1}{2} \left[\rho_1 \dot{u}_1^2 + \rho_2 \dot{u}_2^2 \right] dX, \tag{3.62}$$

where the linear mass densities ρ_1 and ρ_2 are defined through

$$\rho_1 = \frac{M_1}{\varepsilon}, \qquad \rho_2 = \frac{M_2}{\varepsilon} \tag{3.63}$$

The derivation of the internal energy $\mathcal{U}_{int}$ needs the adoption of the following Taylor's series expansion of $u_2(X_{i+1}, t)$ centered at $X = X_i$

$$u^2_{i+1}(t) = u_2(X_{i+1}, t) = u_2(X_i + \varepsilon, t) \cong u_2(X_i, t) + \varepsilon u_2'(X_i, t) + \frac{1}{2}\varepsilon^2\, u_2''(X_i, t). \tag{3.64}$$

which, together with (3.61), leads us to

$$\begin{aligned}\mathcal{U}_{int} &= \frac{\varepsilon}{2}\sum_{i=0}^{N} \frac{K_1}{\varepsilon}\,(u_1(X_i, t) - u_2(X_i, t))^2 + \frac{K_2}{\varepsilon}\left(u_2(X_i, t) + \varepsilon u_2'(X_i, t) - u_1(X_i, t)\right)^2 \\ &= \int_0^L \left[\frac{1}{2}K_N^1\,(u_1 - u_2)^2 + \frac{1}{2}K_N^2\left(u_2 + \varepsilon u_2' - u_1\right)^2\right] dX \end{aligned} \tag{3.65}$$

Here, we have introduced the axial stiffness coefficients K_N^1 and K_N^2 of the continuum diatomic medium as follows

$$K_N^1 = K_1\,\varepsilon, \qquad K_N^2 = K_2\,\varepsilon. \tag{3.66}$$

For the sake of simplicity, and since we are mainly interested to the dispersion relation of the current model, we assume that no external forces are applied to the chain. It would not be difficult, however, to derive the external forces of the continuous diatomic chain, by generalizing the results presented in the previous section. The Action functional is always defined through (3.9), assuming the simplifying hypothesis $\mathcal{U}_{ext} = 0$, and the usual action principle that follows is used to derive the equations of motion in variational form

$$\delta\mathcal{A} = 0, \qquad \forall\delta u_1 \in \mathcal{C}_1, \qquad \forall\delta u_2 \in \mathcal{C}_2, \tag{3.67}$$

for any admissible variations $\mathcal{C}_1$ of u_1 and $\mathcal{C}_2$ of u_2. The Piola's ansatz (3.61) and Eq. (3.28) require

$$u_1(X, t_1) = u_1^1, \quad u_1(X, t_2) = u_2^1, \quad u_2(X, t_1) = u_2^1, \quad u_2(X, t_2) = u_2^2, \tag{3.68}$$

It is easily found that the first variation of the action functional is written as follows in the present case

$$\begin{aligned}\delta\mathcal{A} &= \int_{t_1}^{t_2}\left[\int_0^L \left[\delta u_1\left(K_N^1(u_1 - u_2) - K_N^2\left(u_2 - u_1 + \varepsilon u_2'\right) + \rho_1\ddot{u}_1\right)\right] dX\right] dt \\ &+ \int_{t_1}^{t_2}\left[\int_0^L \left[\delta u_2\left(-K_N^1(u_1 - u_2) + K_N^2\left(u_2 - u_1\right) - \varepsilon K_N^2(\varepsilon u_2'' - u_1') + \rho_2\ddot{u}_2\right)\right] dX\right] dt \\ &+ \int_{t_1}^{t_2}\left[\varepsilon\delta u_2\left(K_N^2\left(u^2 + \varepsilon u^{2\prime} - u^1\right)\right)\right]_0^L dt \end{aligned} \tag{3.69}$$

Invoking the arbitrariness of the variations in (3.67), we obtain the following PDEs

$$K_N^1(u_1 - u_2) - K_N^2\left(u_2 - u_1 + \varepsilon u_2'\right) + \rho_1 \ddot{u}_1 = 0 \tag{3.70}$$

$$-K_N^1(u_1 - u_2) + K_N^2\left(u_2 - u_1\right) - \varepsilon K_N^2(\varepsilon u_2'' - u_1') + \rho_2 \ddot{u}_2 = 0 \tag{3.71}$$

Assuming now plane wave solutions for the displacement fields $u_1(X, t)$ and $u_2(X, t)$ defined by

$$u_1 = \mathrm{Re}\left\{u_0^1 \exp\left[I\left(\omega t - k_w X\right)\right]\right\}, \quad u_2 = \mathrm{Re}\left\{u_0^2 \exp\left[I\left(\omega t - k_w X\right)\right]\right\}, \tag{3.72}$$

where u_0^1, u_0^2, ω and k_w are, respectively, the complex amplitudes, the angular frequency and the wave number, we finally obtain the following dispersion relation of the continuum model of the diatomic chain

$$\omega_{Doc,Dac}\left(k_w\right) = \sqrt{\frac{f_D(k_w) \pm \Delta_D\left(k_w\right)}{\rho_1 \rho_2}} \tag{3.73}$$

where the function $f_D(k_w)$ and the discriminant $\Delta_D\left(k_w\right)$ are written as

$$f_D(k_w) = (\rho_1 + \rho_2)(K_N^1 + K_N^2) + \rho_1 K_N^2 k_w^2 \varepsilon^2 \tag{3.74}$$

$$\Delta_D\left(k_w\right) = \sqrt{[f_D(k_w)]^2 - 4\rho_1\rho_2 K_N^1 K_N^2 k_w^2 \varepsilon^2} \tag{3.75}$$

It is worth to note that, in the limit for $\varepsilon \to 0$, one obtains

$$\omega_{Dac}\left(k_w\right) \cong k_w \sqrt{\frac{K_N^1 K_N^2}{(K_N^1 + K_N^2)(\rho_1 + \rho_2)}}, \tag{3.76}$$

$$\omega_{Doc}\left(k_w\right) \cong \sqrt{\frac{(K_N^1 + K_N^2)(\rho_1 + \rho_2)}{\rho_1 \rho_2}}, \tag{3.77}$$

Such equations, in force of the identification formulae (3.63) and (3.66), correctly reduce to Eqs. (3.35) and (3.36) of the discrete model. Figure 3.9 shows a graphical illustration of the dispersion relation (3.73) for the continuum model of the diatomic chain #1 studied in Sect. 3.3.2. It is seen that the acoustic and optic branches of such a relation accurately reproduce the analogous branches of the discrete dispersion relation, as $k_w \to 0$.

3.5 Concluding Remarks

We have presented discrete and continuum models of monoatomic and diatomic mass-spring chains with tensegrity architecture. Such models significantly generalize previous ones available in the literature (see, e.g., Amendola et al. 2018; Krushynska et al. 2018), due to combination of prism units with different architectures, the modeling of the coupling between axial and rotational motions of the masses forming

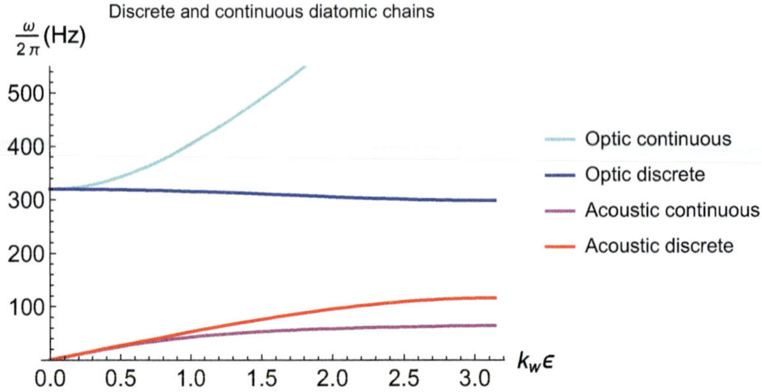

Fig. 3.9 Comparison between the dispersion relations of discrete and continuous models of the diatomic chain #1 analyzed in Sect. 3.3.2. In the long wavelength limit ($k_w \to 0$), it is seen that the acoustic and optic branches of the dispersion relation of the continuous model well reproduce the analogous branches of the relative dispersion relations of the discrete model

the analyzed chains, and the formulation of second-gradient, continuum limits of the discrete chain models in the long wavelength regime. Numerical results dealing with physical models of monoatomic and diatomic chains, which are equipped with micro-scale prisms, have shown that it is possible to finely tune the bandgap regions by acting on a large number of design variables. These include geometry, prestress and topology variables; the material and the size of the lumped masses, as well as the possibility to couple or uncouple the rotational motions of the prisms and the lumped masses. The extreme tunability properties of bandgap metamaterials with tensegrity architecture pave the way to their employment to design novel, exciting acoustic systems, like, e.g., vibration protection and shock protector devices; acoustic lenses; and energy trapping containers (refer, e.g., to Theocharis et al. 2013 and references therein).

Future extensions of the present study will analyze the dispersive nature of the wave dynamics of tensegrity lattices in the large displacement regime, the design and the mechanical modeling of tensegrity metamaterials in multiple dimensions, the inclusion of internal resonance effects (Placidi et al., 2024b); Placidi et al. (2024a), as well as experimental validations of the proposed models, by recourse to dynamic tests on prototypes fabricated through additive manufacturing at different scales (Amendola et al., 2014; Amendola et al., 2015; Pajunen et al., 2019; Intrigila et al., 2023; Al Sabouni-Zawadzka et al., 2024). Additional lines of future research could explore the use of tensegrity bandgap systems in the design of seismic lattice metamaterials (Fraternali et al. 2018, 2021; Santos et al., 2022).

References

Al Sabouni-Zawadzka, A., Gilewski, W., Nazifi Charandabi, R., & Zawadzki, A. (2024). Stability of tensegrity-inspired structures fabricated through additive manufacturing. *Composite Structures, 345*, 118377.

Amendola, A., Hernández-Nava, E., Goodall, R., Todd, I., Skelton, R. E., & Fraternali, F. (2015). On the additive manufacturing, post-tensioning and testing of bi-material tensegrity structures. *Composite Structures, 131*, 66–71.

Amendola, A. (2023). An analytic study on the properties of solitary waves traveling on tensegrity-like lattices. *International Journal of Non-Linear Mechanics, 148*, 104264.

Amendola, A., Carpentieri, G., de Oliveira, M., Skelton, R. E., & Fraternali, F. (2014). Experimental investigation of the softening-stiffening response of tensegrity prisms under compressive loading. *Composite Structures, 117*, 234–243.

Amendola, A., Krushynska, A., Daraio, C., Pugno, N. M., & Fraternali, F. (2018). Tuning frequency band gaps of tensegrity mass-spring chains with local and global prestress. *International Journal of Solids and Structures, 155*, 47–56. ISSN 0020-7683.

Ashcroft, N. W., & David Mermin, N. (1976). *Solid state physics*. New York: Holt, Rinehart and Winston.

Barchiesi, E., Eugster, S. R., Placidi, L., & dell'Isola, F. (2019). Pantographic beam: a complete second gradient 1d-continuum in plane. *Zeitschrift für angewandte Mathematik und Physik, 70*, 1–24.

Barchiesi, E., & Khakalo, S. (2019). Variational asymptotic homogenization of beam-like square lattice structures. *Mathematics and Mechanics of Solids, 24*(10), 3295–3318.

de Castro Motta, J., Placidi, L., Nazifi Charandabi, R., & Fraternali, F. (2024). A Weierstrass approach to the analysis of rarefaction solitary waves in tensegrity mass-spring systems. *Communications in Applied and Industrial Mathematics, 15*(2), 8–16.

Fraternali, F., Senatore, L., & Daraio, C. (2012). Solitary waves on tensegrity lattices. *Journal of the Mechanics and Physics of Solids, 60*(6), 1137–1144.

Fraternali, F., Carpentieri, G., Amendola, A., Skelton, R. E., & Nesterenko, V. F. (2014). Multi-scale tunability of solitary wave dynamics in tensegrity metamaterials. *Applied Physics Letters, 105*(20).

Fraternali, F., Carpentieri, G., & Amendola, A. (2015). On the mechanical modeling of the extreme softening/stiffening response of axially loaded tensegrity prisms. *Journal of the Mechanics and Physics of Solids, 74*, 136–157.

Fraternali, F., Amendola, A., & Benzoni, G. (2018). Innovative seismic isolation devices based on lattice materials: A review. *Ingegneria Sismica/International Journal of Earthquake Engineering, 35*(4), 93–113.

Fraternali, F., Singh, N., Amendola, A., Benzoni, G., & Milton, G. W. (2021). A biomimetic sliding-stretching approach to seismic isolation. *Nonlinear Dynamics, 106*(4), 3147–3159.

Huber, S. (2018). Mechanical metamaterials lecture notes. www.cmt-qo.phys.ethz.ch

Intrigila, C., Micheletti, A., Nodargi, N. A., & Bisegna, P. (2023). Mechanical response of multistable tensegrity-like lattice chains. *Additive Manufacturing, 74*, 103724.

Kadic, M., Milton, G. W., van Hecke, M., & Wegener, M. (2019). 3d metamaterials. *Nature Reviews Physics, 1*(3), 198–210.

Krushynska, A. O., Amendola, A., Bosia, F., Daraio, C., Pugno, N. M., & Fraternali, F. (2018). Accordion-like metamaterials with tunable ultra-wide low-frequency band gaps. *New Journal of Physics, 20*(7), 073051.

Landau, L. D., & Lifshitz, E. M. (1976). *Mechanics, Third Edition: Volume 1 (Course of Theoretical Physics)*. Butterworth-Heinemann.

Liu, Z., Zhang, X., Mao, Y., Zhu, Y. Y., Yang, Z., Chan, C. T., & Sheng, P. (2000). Locally resonant sonic materials. *Science, 289*(5485), 1734–1736.

Lu, M.-H., Feng, L., & Chen, Y.-F. (2009). Phononic crystals and acoustic metamaterials. *Materials Today, 12*(12), 34–42. ISSN 1369-7021.

Mascolo, I., Amendola, A., Zuccaro, G., Feo, L., & Fraternali, F. (2018). On the geometrically nonlinear elastic response of class θ=1 tensegrity prisms. *Frontiers in Materials, 5*.

Micheletti, A., Ruscica, G., & Fraternali, F. (2019). On the compact wave dynamics of tensegrity beams in multiple dimensions. *Nonlinear Dynamics, 98*(4), 2737–2753.

Pajunen, K., Johanns, P., Pal, R. K., Rimoli, J. J., & Daraio, C. (2019). Design and impact response of 3d-printable tensegrity-inspired structures. *Materials & Design, 182*, 107966.

Placidi, L., de Castro Motta, J., & Fraternali, F. (2024). Bandgap structure of tensegrity mass-spring chains equipped with internal resonators. *Mechanics Research Communications, 137*, 104273.

Placidi, L., de Castro Motta, J., Nazifi Charandabi, R., & Fraternali, F. (2024). A continuum model for the tensegrity maxwell chain. *NHM, 19*(2), 597–610.

Santos, F. A., Caroço, C., Amendola, A., Miniaci, M., & Fraternali, F. (2022). 3D tensegrity braces with superelastic response for seismic control. *International Journal for Multiscale Computational Engineering, 20*(5), 53–64.

Skelton, R. E., & de Oliveira, M. C. (2009). *Tensegrity systems* (Vol. 1). Berlin: Springer.

Theocharis, G., Boechler, N., & Daraio, C. (2013). Nonlinear phononic structures and metamaterials. In *Acoustic matematerials and phononic crystals, Springer series in solid state sciences* (Vol. 173).

Turco, E., Misra, A., Pawlikowski, M., dell'Isola, F., & Hild, F. (2018). Enhanced piola-hencky discrete models for pantographic sheets with pivots without deformation energy: Numerics and experiments. *International Journal of Solids and Structures, 147*, 94–109.

Turco, E., Barchiesi, E., Giorgio, I., & dell'Isola, F. (2020). A lagrangian hencky-type non-linear model suitable for metamaterials design of shearable and extensible slender deformable bodies alternative to timoshenko theory. *International Journal of Non-Linear Mechanics, 123*, 103481.

Chapter 4
Propagation of Solitary Waves in Geometrically Nonlinear Tensegrity Lattices

Fernando Fraternali, Julia de Castro Motta, and Ada Amendola

Abstract The propagation of compression and rarefaction solitary waves in tensegrity mass-spring chains under impact loading is analyzed using analytical methods. The Weierstrass criterion is applied to detect solitary wave solutions in the continuum limit of the examined system. The study utilizes an interaction potential that preserves the fundamental characteristics observed in minimal and $\theta = 1$ tensegrity prisms under compression loading. The Chapter includes a review of literature results on the propagation of solitary waves in 1D, 2D, and 3D tensegrity lattices. The analyzed wave regime offers potential applications for designing novel acoustic lenses with tunable focus, innovative actuators and sensors for nondestructive structural health monitoring, and impact protection devices.

Keywords Solitary waves · Weierstrass criterion · Tensegrity-like potential · Waveform

4.1 Introduction

Several recent studies have shown that the geometrically nonlinear response of tensegrity mass-spring systems supports the propagation of solitary waves under impact loading. These waves range from compact compression waves (Fraternali et al., 2012; Micheletti et al., 2019; Amendola, 2023a; de Castro Motta et al., 2024b) to rarefaction solitary waves (Fraternali et al., 2014; de Castro Motta et al., 2024c), depending on whether the unit response switches from stiffening to softening, respectively. Solitary wave dynamics have been proposed for constructing various novel

F. Fraternali (✉) · J. de Castro Motta · A. Amendola
Department of Civil Engineering, University of Salerno, Via Giovanni Paolo II 132, 84084 Fisciano, SA, Italy
e-mail: f.fraternali@unisa.it

J. de Castro Motta
e-mail: jdecastromotta@unisa.it

A. Amendola
e-mail: adaamendola1@unisa.it

F. Fraternali and J. J. Rimoli (eds.), *Tensegrity Systems*, CISM International Centre for Mechanical Sciences 2, https://doi.org/10.1007/978-3-031-82283-4_4

acoustic devices, such as impact mitigation systems and tunable acoustic lenses. Effective impact mitigation systems based on tensegrity structures with a softening-type response can transform compressive disturbances into solitary rarefaction waves with a progressively vanishing oscillatory tail (Fraternali et al., 2015; Herbold & Nesterenko, 2013). Tunable acoustic lenses made from elements with a stiffening response can spatially focus compression solitary waves in different regions (Fraternali et al., 2012; Spadoni & Daraio, 2010).

This Chapter complements the one focused on the (linear) bandgap behavior of tensegrity mass-spring chains by exploring the solitary wave dynamics exhibited by these systems in the presence of a geometrically nonlinear response regime characterized by large strains of the cables and either elastically stiffening or elastically softening behavior. The examined tensegrity units consist of regular minimal (T3) prisms and $\theta = 1$ (Tθ) prisms. Due to the complex behavior of the force versus displacement response in the compression of these prisms, it is not feasible to obtain analytical results for the wave propagation response exhibited by mass-spring systems that use such units alternated with lumped masses. The use of a 'tensegrity-like' potential that mimics the nature of the unit cell's potential has been successfully applied to analytic studies of solitary wave propagation, both in pre-compressed (de Castro Motta et al., 2024c) and non-pre-compressed chains of T3 prisms (Amendola, 2023a), as well as truncated octahedron tensegrities (de Castro Motta et al., 2024b).

A tensegrity-like interaction potential is hereafter used to demonstrate the existence of solitary waves in chains composed of two different prisms examined in the Bandgap Chapter (T3D and TθD), alternated with lumped masses. The focus is on two different responses: elastically softening and elastically stiffening. The lower and upper bounds of the wave speed interval in which the solitary waves propagate is provided analytically, along with the analytic profiles of the solitary pulses. The structure of the Chapter is as follows. Section 4.2 reviews a tensegrity-like interaction potential and its main features. Next, a fitting procedure is employed to simulate the responses of the examined tensegrity units with tensegrity-like models. In Sect. 4.3, the quasi-continuum limit of the equations of motion of a tensegrity-like mass-spring chain is derived. A Weierstrass analysis of these wave equations is presented in Sect. 4.4, demonstrating the existence of compression and rarefaction solitary waves in stiffening and softening systems, respectively, and outlining their key features. Section 4.5 provides numerical results for systems equipped with tensegrity-like potentials and reviews available numerical findings from the literature on the propagation of compression and rarefaction waves in 1D, 2D, and 3D tensegrity lattices (Fraternali et al., 2014; Micheletti et al., 2019). These results qualitatively validate the theoretical analysis presented in Sect. 4.4, as they demonstrate the presence of compression and rarefaction pulses in physical examples of tensegrity lattices across multiple dimensions. Quantitative comparisons between the predictions of tensegrity-like models and numerical simulations from the literature are provided in Amendola (2023a), de Castro Motta et al. (2024a). The Chapter concludes in Sect. 4.6 with final remarks and suggestions for future research directions.

4.2 Tensegrity-Like Interaction Potential

We present hereafter a non-linear tensegrity-like interaction potential, which is aimed at capturing the main features of the mechanical response of T3D and TθD prisms. The presented model has proven to be suitable for mimicking tensegrity units with stiffening or softening response in previous work (de Castro Motta et al., 2024b, 2024c). Its main property lies in the fact that it predicts axial stiffness tending to a constant value for sufficiently large relative axial displacements between the terminal bases of the tensegrity prism.

Let h, h_0, and $\hat{h}_0$ respectively denote the height of the tensegrity prism unit in the current (deformed) configuration, the prism height in the self-standing configuration (i.e., the configuration under zero external forces and the action of internal, self-equilibrated prestress), and the prism height in the externally pre-compressed configuration, which is obtained by applying compressive forces with amplitude F_0 to the centers of mass of the terminal bases. The height h is given by $h = h_0 - \Delta\hat{u} = \hat{h}_0 - \Delta u$, where $\Delta\hat{u}$ denotes the total (relative) axial displacement of the prism from the self-standing configuration (i.e., the relative axial displacement between the terminal bases), and Δu denotes the incremental (relative) displacement from the externally pre-compressed configuration. All such displacements are assumed to be positive in compression. The displacement taking the prism from the self-standing to the externally pre-compressed configuration will be denoted by $\Delta\hat{u}_{F_0}$ in the sequel ($\Delta\hat{u}_{F_0} = \hat{h}_0 - h_0 = \Delta\hat{u} - \Delta u$).

We now consider the tensegrity-like interaction potential law V versus Δu introduced in de Castro Motta et al. (2024b), which can be cast into the following form

$$V(\Delta u) = \alpha_1 \, \Delta u^2 \, \frac{\alpha_2 + \Delta u}{\alpha_3 + \Delta u} \tag{4.1}$$

where α_1, α_2 and α_3 are constitutive parameters. By taking derivatives we obtain the incremental axial force F versus Δu and tangent stiffness k versus Δu laws, which are given by

$$F(\Delta u) = \frac{dV}{d\Delta u} = \alpha_1 \Delta u \frac{2\alpha_2\alpha_3 + (\alpha_2 + 3\alpha_3)\,\Delta u + 2\Delta u^2}{(\alpha_3 + \Delta u)^2} \tag{4.2}$$

$$k(\Delta u) = \frac{dF}{d\Delta u} = 2\alpha_1 \frac{\alpha_2\alpha_3^2 + 3\alpha_3\Delta u^2 + 3\alpha_3^2\Delta u + \Delta u^3}{(\alpha_3 + \Delta u)^3} \tag{4.3}$$

Let the (initial) tangent stiffness for $\Delta u = 0$ be denoted by k_0. Using Eq. (4.3) we obtain

$$k_0 = 2\,\frac{\alpha_1\alpha_2}{\alpha_3}. \tag{4.4}$$

It is worth noting that the $k(\Delta u)$ law defined by Eq. (4.3) tends to an horizontal asymptote $k = k_f$ for $\Delta u \rightarrow \infty$, with k_f given by

$$k_f = 2\,\alpha_1 = \frac{\alpha_3}{\alpha_2}\,k_0 \tag{4.5}$$

From Eq. (4.5), we deduce that $\alpha_1 > 0$ is necessary to predict a strictly positive value of k_f. Using this result, we further deduce from Eq. (4.4) that $\alpha_2 \geq 0$ is required to predict $k_0 \geq 0$. Notably, $k_0 = 0$ is obtained for $\alpha_2 = 0$, which is useful for describing the response of a prism under zero pre-compression (see, e.g., Fraternali et al., 2014). Our subsequent analysis needs to determine the values of the α_2/α_3 ratio that can model stiffening- and softening-type interaction potentials, respectively. According to Eq. (4.3), an elastically stiffening response (i.e., a response that features k increasing with increasing values of Δu) is obtained when $0 < \alpha_2/\alpha_3 < 1$. Conversely, an elastically softening response (k decreasing with increasing values of Δu) is obtained when $0 < \alpha_3/\alpha_2 < 1$. Figure 4.1a, b, c visually illustrate the V versus Δu, F versus Δu, and k versus Δu laws, derived from Eqs. (4.1)–(4.3) for specific values of the constitutive parameters α_1, α_2, and α_3. These figures show both stiffening (orange and green curves) and softening (magenta curves) types of response.

4.2.1 Fitting the Tensegrity-Like Potential to the Response of Tensegrity Prisms

We now move on to demonstrate that the tensegrity-like model accurately describes the behavior of T3D and TθD prisms presented in the Bandgap Chapter. For the T3D prism, we examine the incremental response from a slightly externally pre-compressed configuration with $F_0 = 1.11$ N and $\Delta\hat{u}_{F_0} = 0.092$ mm (a configuration close to the origin of the plot in Fig. 3.1d of the Bandgap Chapter). During this response, the displacement from the self-standing configuration grows up to $\Delta\hat{u} = 0.909$ mm, allowing us to capture the stiffening branch of the response shown in Fig. 3.1d of the Bandgap Chapter. For the TθD prism, we study the incremental response from a strongly externally pre-compressed configuration with $F_0 = 76.716$ N and $\Delta\hat{u}_{F_0} = 1.603$ mm (the final point of the curve shown in Fig. 3.2d of the Bandgap Chapter). The examined incremental response unloads the TθD prism from the pre-compressed state back to a configuration with zero total displacement $\Delta\hat{u}$, enabling us to capture the softening-type response illustrated in Fig. 3.2d of the Bandgap Chapter. We will see later that these choices lead us to predict the propagation of compression solitary waves in mass-spring systems equipped with T3D units and rarefaction solitary waves in mass-spring systems equipped with TθD units.

Figure 4.2a shows the fitting of the tensegrity-like model to the examined response branch of the T3D prism. This result was obtained using a set of 967 fitting points uniformly spaced along the Δu-axis for $\Delta u \in [0, 0.817]$ mm. We used the symbol

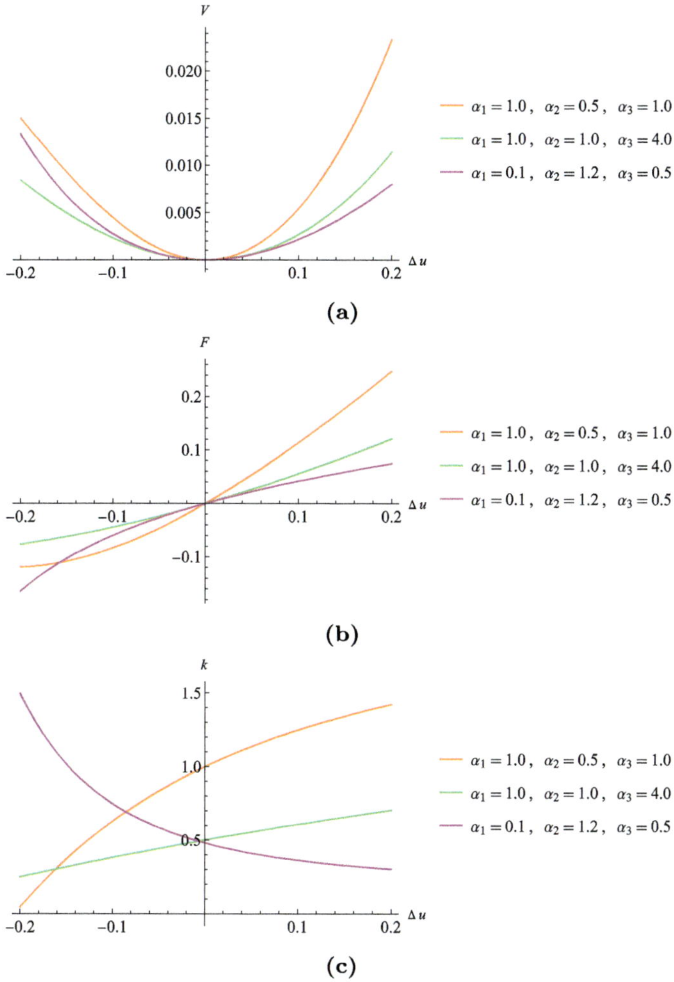

Fig. 4.1 V versus Δu (**a**), F versus Δu (**b**), and k versus Δu (**b**) laws for selected values of the constitutive parameters α_1, α_2 and α_3

F_{T3D} to denote the actual response of the T3D unit, and the symbol F to denote the tensegrity-like response. The Mathematica® function FindFit determined the best-fit constitutive parameters: $\alpha_1 = 1.73596 \times 10^7$ N/m, $\alpha_2 = 2.57527 \times 10^{-3}$ m, and $\alpha_3 = 7.86721$ m. The results in Fig. 4.2a demonstrate a rather good match between the tensegrity-like model and the T3D unit response, in correspondence to the examined response branch.

The fitting of the tensegrity-like model to the studied response branch of the TθD prism is illustrated in in Fig. 4.2b. In this case, we used 97 fitting points uniformly spaced along the Δu-axis for $\Delta u \in [-1.603, 0]$ mm, which lead us to obtain: $\alpha_1 = 7.32552$ N/m, $\alpha_2 = 13.7383$ m and $\alpha_3 = 6.45442 \times 10^{-3}$ m. We let $F_{T\theta D}$ denote

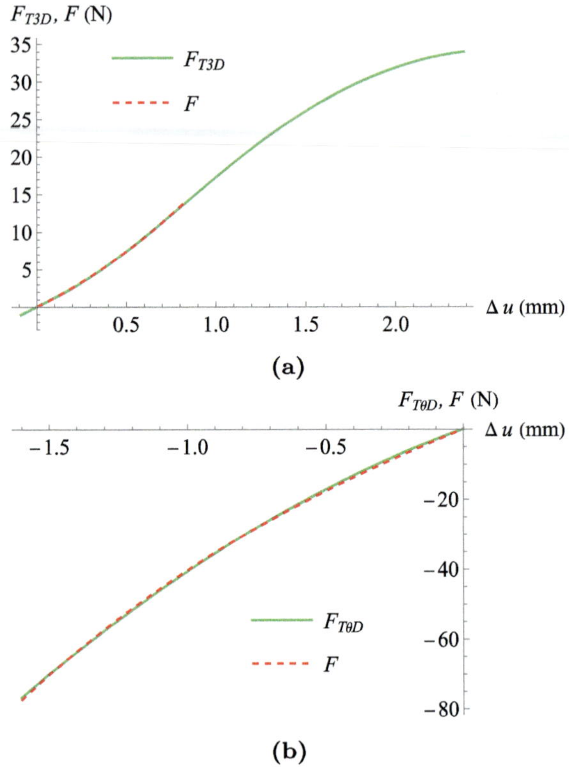

Fig. 4.2 Fitting of the force versus displacements responses of tensegrity units T3D (**a**), and TθD (**b**) (green-solid curves) with the tensegrity-like model (red-dashed curves)

the actual response of the TθD unit, and again we let F denote the tensegrity-like response. The comparison illustrated in Fig. 4.2b highlights also in the present case a very good matching between the tensegrity-like response and that of the TθD prism, similarly to what we observed for the T3D unit.

4.3 Continuum-Limit of a Mass-Spring Chain

We now wish to derive the *quasi-continuum* limit equations of a pre- compressed tensegrity-like mass-spring chain, which is composed of an infinitely large number of units. Introducing a longitudinal coordinate x along the axis of the chain, we let u_i, u_{i+1} and u_{i-1} denote the displacements exhibited by the masses i, $i+1$ and $i-1$ with coordinates x_i, $x_{i+1} = x_i + \hat{h}_0$ and $x_{i-1} = x_i - \hat{h}_0$, respectively. From this point forward, for the sake of simplicity, we will denote $\hat{h}_0$ as simply h_0, even though the distinction between these quantities, introduced in the previous section, should be kept in mind. The equation of motion of the i-the mass in the longitudinal direction is written as follows

$$m\ddot{u}_i = F(u_{i-1} - u_i) - F(u_i - u_{i+1}) \tag{4.6}$$

It is worth noting that the compressive strain of the i-th tensegrity unit is proportional to the displacement $\Delta u_i = u_{i-1} - u_i$. Consequently, we can rewrite Eq. (4.6) as

$$m\Delta\ddot{u}_i = F(\Delta u_{i+1}) - 2F(\Delta u_i) + F(\Delta u_{i-1}). \tag{4.7}$$

We begin by writing the quantity Δu_i as the realization of a continuous function $\Delta u(x, t)$ in correspondence to the current time t and the position x_i of the i-th mass (i.e., $\Delta u_i = \Delta u(x_i, t)$). By using a discrete-to-continuum approach, we next employ fourth-order Taylor series expansions of $F(\Delta u_{i+1})$ and $F(\Delta u_{i-1})$ around $x = x_i$. Such expansions result in

$$F(\Delta u_{i+1}) = F(\Delta u_i) + h_0 F_x(\Delta u_i) + \frac{{h_0}^2}{2} F_{xx}(\Delta u_i) + \frac{{h_0}^3}{6} F_{xxx}(\Delta u_i) + \frac{{h_0}^4}{24} F_{xxxx}(\Delta u_i)$$

and

$$F(\Delta u_{i-1}) = F(\Delta u_i) - h_0 F_x(\Delta u_i) + \frac{{h_0}^2}{2} F_{xx}(\Delta u_i) - \frac{{h_0}^3}{6} F_{xxx}(\Delta u_i) + \frac{{h_0}^4}{24} F_{xxxx}(\Delta u_i)$$

Inserting the above approximations into Eq. (4.7), we obtain

$$m\Delta u_{tt} = {h_0}^2 F_{xx}(\Delta u) + \frac{{h_0}^4}{12} F_{xxxx}(\Delta u)$$

Recalling that $F = V_{\Delta u}$, we can rewrite the last equation as

$$\frac{m}{{h_0}^2}\Delta u_{tt} = \mathfrak{L}\{[V_{\Delta u}(\Delta u(x, t))]_{xx}\} \tag{4.8}$$

where

$$\mathfrak{L} \approx 1 + \frac{1}{12}{h_0}^2 \partial_{xx}$$

is a differential operator.

Assuming that Δu is small compared to h_0 ($\Delta u \sim \epsilon\Delta u$, where $\epsilon \ll 1$), but not an infinitesimal of the first order, we can perform a Taylor expansion of $V(\Delta u)$ around $\epsilon\Delta u = 0$ in terms of ϵ. Given that $V(0) = F(0) = 0$, this expansion can be written as follows

$$V(u) \approx \frac{1}{2}k(0)\epsilon^2 \Delta u^2 + \frac{1}{6}k'(0)\epsilon^3 \Delta u^3 + \dots \tag{4.9}$$

We follow Rosenau's approach (Rosenau, 1986) to invert the operator $\mathfrak{L}$, obtaining

$$\mathfrak{L}^{-1} \approx \left(1 - \frac{1}{12}{h_0}^2 \partial_{xx}\right)$$

Using this inverse operator into Eq. (4.8), we can derive the modified Boussinesq Equation given by (de Castro Motta et al., 2024c)

$$\Delta u_{tt} = \frac{{h_0}^2}{m}\left[V_{\Delta u}\right]_{xx} + \gamma \Delta u_{xxtt} \tag{4.10}$$

where

$$\gamma = \frac{{h_0}^2}{12}.$$

It is useful to introduce the following modified potential, which is a function of the dimensionless variable $\xi = \Delta u / h_0$

$$\tilde{V}(\xi) = \frac{V(h_0 \xi)}{m} = \tilde{\alpha}_1 \, \xi^2 \frac{\tilde{\alpha}_2 + \xi}{\tilde{\alpha}_3 + \xi} \tag{4.11}$$

where

$$\tilde{\alpha}_1 = \frac{\alpha_1 {h_0}^2}{m}; \quad \tilde{\alpha}_2 = \frac{\alpha_2}{h_0}; \quad \tilde{\alpha}_3 = \frac{\alpha_3}{h_0}. \tag{4.12}$$

A simple manipulation of Eq. (4.10), using Eq. (4.11), leads us to the following differential equation

$$\xi_{tt} = \left[\tilde{V}_\xi\right]_{xx} + \gamma \xi_{xxtt} \tag{4.13}$$

Let us now consider a strain wave ξ traveling with speed $v \geq 0$. We express the traveling pulse as $\xi = \phi(x - vt)$, where ϕ is a scalar function to be determined. By setting $z = x - vt$, we can rewrite Eq. (4.13) as follows

$$v^2 \phi'' = \left[\tilde{V}_\phi\right]'' + v^2 \gamma \phi'''' \tag{4.14}$$

having used the prime notation for the differentiation with respect to z. Integrating twice with respect to z and employing asymptotic boundary conditions to remove integration constants results in the transformation of Eq. (4.14) into the form

$$v^2 \gamma \phi'' = v^2 \phi - \left[\tilde{V}_\phi\right]. \tag{4.15}$$

Multiplying Eq. (4.15) by ϕ' on both sides and once more integrating with respect to z, we finally obtain

$$\mathcal{F} = \frac{1}{v^2}\left(v^2\phi^2 - 2\tilde{V} + C\right), \tag{4.16}$$

where C is an integration constant, and we have set

$$\mathcal{F} = \gamma\, \phi'^2. \tag{4.17}$$

As ϕ' is a real function and $\gamma > 0$, it is worth noting that it must result in $\mathcal{F} \geq 0$.

4.4 Weierstrass Analysis of Solitary Waves

The present Section is dedicated to carry out a Weierstrass analysis of Eq. (4.16) (refer, e.g., to de Castro Motta et al., 2024c and references therein). It is known (Saccomandi, 2004; Destrade et al., 2007) that solitary waves will exist when Eq. (4.16) presents a double zero (corresponding to an asymptotic point) and a simple zero (corresponding to an inversion point).

Inserting Eq. (4.11) in (4.16), we can write $\mathcal{F}$ as

$$\mathcal{F} = \phi^2\left[1 - \frac{2}{v^2}\,\tilde{\alpha}_1\frac{\tilde{\alpha}_2 + \phi}{\tilde{\alpha}_3 + \phi}\right]. \tag{4.18}$$

It is clear that $\phi = 0$ is a double root of Eq. (4.18), corresponding to an asymptotic point of the wave profile. In contrast, ϕ^* is a simple root of $\mathcal{F} = 0$, which corresponds to an inversion point of the wave profile. We can express ϕ^* from Eq. (4.18) as follows

$$\phi^* = \frac{\tilde{\alpha}_3 v^2 - 2\tilde{\alpha}_1\tilde{\alpha}_2}{2\tilde{\alpha}_1 - v^2}. \tag{4.19}$$

It is useful to introduce a reference limit displacement, which we shall define as $\Delta u_{lim} = h_0\,\xi_{lim}$. This value can be defined by several events, such as the buckling of the bars, the locking of the structure, or a point where the prism significantly changes its behavior. In order to guarantee that $|\phi^*| \in \left]0,\ |\xi_{lim}|\right]$ (i.e., $\mathcal{F} > 0$), we have to identify the maximum and minimum speeds that suit these conditions. By setting $\phi^* = 0$ into (4.19), we obtain the minimum speed into the following form

$$v_{min} = \sqrt{2\tilde{\alpha}_1\frac{\tilde{\alpha}_2}{\tilde{\alpha}_3}} = c_0 \tag{4.20}$$

and we observe that v_{min} coincides with the speed of sound in the linearized system $c_0 = h_0\sqrt{\frac{k_0}{m}}$.

Similarly, enforcing $|\phi^*| < |\xi_{lim}|$ in Eq. (4.19), we find the following expression of the maximum wave speed

$$v_{max} = \sqrt{2\tilde{\alpha}_1 \frac{\tilde{\alpha}_2 + \xi_{lim}}{\tilde{\alpha}_3 + \xi_{lim}}}. \tag{4.21}$$

It is easily observed that a stiffening response ($\alpha_2 < \alpha_3$) with $v_{min} < v < v_{max}$ results in $0 < \phi^* \le \xi_{lim}$, which corresponds to a compression-type solitary pulse. Conversely, in the case of a softening response ($\alpha_2 > \alpha_3$) with $v_{min} < v < v_{max}$, one obtains $\xi_{lim} \le \phi^* < 0$, implying that the solitary wave is a rarefaction-type pulse.

From Eq. (4.17), we know that $\frac{d\phi}{dz} = \sqrt{\frac{\mathcal{F}}{\gamma}}$. Introducing $f = \pm z$, we can perform the implicit integration

$$f := \int \sqrt{\frac{\gamma}{\mathcal{F}}}\, d\phi = \pm z. \tag{4.22}$$

Inserting $\mathcal{F}$ from Eq. (4.18) in Eq. (4.22), we obtain

$$f = 2\, v \sqrt{\gamma}\; \frac{\psi \tan^{-1}\left(\frac{\chi}{\rho}\right) - \rho \tan^{-1}\left(\frac{\chi}{\psi}\right)}{\psi \rho} \tag{4.23}$$

where

$$\chi = \sqrt{v^2 - \frac{2\tilde{V}}{\phi^2}}, \qquad \rho = \sqrt{c_0{}^2 - v^2}, \qquad \psi = \sqrt{2\tilde{\alpha}_1 - v^2}. \tag{4.24}$$

4.5 Numerical Results

4.5.1 Illustration of Analytic Results

Let us investigate the nature of the waves traveling in the continuum-limit of mass-spring systems exhibiting the tensegrity-like potentials illustrated in Sect. 4.2.1. This will enable us to determine the wave speeds at which compression and rarefaction pulses can propagate in these systems, as well as the shape of such pulses. For all results presented hereafter, we assume that the lumped masses forming the mass-spring chain each weigh 25 g.

Using the best-fit constitutive parameters for the T3D unit ($\alpha_1 = 1.73596 \times 10^7$ N/m, $\alpha_2 = 2.57527 \times 10^{-3}$ m, $\alpha_3 = 7.86721$ m), Eq. (4.20) shows that the minimum speed at which a solitary pulse can propagate is $v_{min} = 3.62282$ m/s. To ensure the units remain in the stiffening phase, we assume a limit value $\Delta u_{lim} = 0.817$ mm, corresponding to $\xi_{lim} = 0.151$. According to Eq. (4.21), the maximum speed is

$v_{max} = 4.15893$ m/s. We can conclude that mass-spring systems equipped with T3D units support compression solitary pulses, as ϕ^* is positive for values of v between v_{min} and v_{max}. This conclusion is confirmed by analyzing the shape of such a pulse (Fig. 4.3a), derived using Eq. (4.23) and the transformation of the axis $f(\phi) \rightarrow \phi(f)$. It is worth noting that the horizontal axis f/h_0 in the plots of Fig. 4.3a corresponds to the number of units traversed by the pulse in the discrete system.

We now study the waves traversing a mass-spring system equipped with TθD units, using the best-fit parameters $\alpha_1 = 7.32552$ N/m, $\alpha_2 = 13.7383$ m, and $\alpha_3 = 6.45442 \times 10^{-3}$ m, which model an elastically softening behavior of the generic unit. To always guarantee a non-zero level of compression in the chain, we set $\Delta u_{lim} \leq \Delta u < 0$, with $\Delta u_{lim} = -1.603$ mm ($\xi_{lim} = -0.171$). Using Eqs. (4.20) and (4.21), we find that $v_{min} = 23.4669$ m/s and $v_{max} = 27.0673$ m/s in this case. Additionally, using Eq. (4.23), we plot the shape of the solitary pulse as illustrated in Fig. 4.3b. One observes that the softening-type system under examination supports the propagation of rarefaction pulses characterized by negative values of ϕ, which progressively reduce the initial state of pre-compression of the chain. It is worth noting that in both plots shown in Fig. 4.3, the width of the solitary pulse decreases as the wave speed increases.

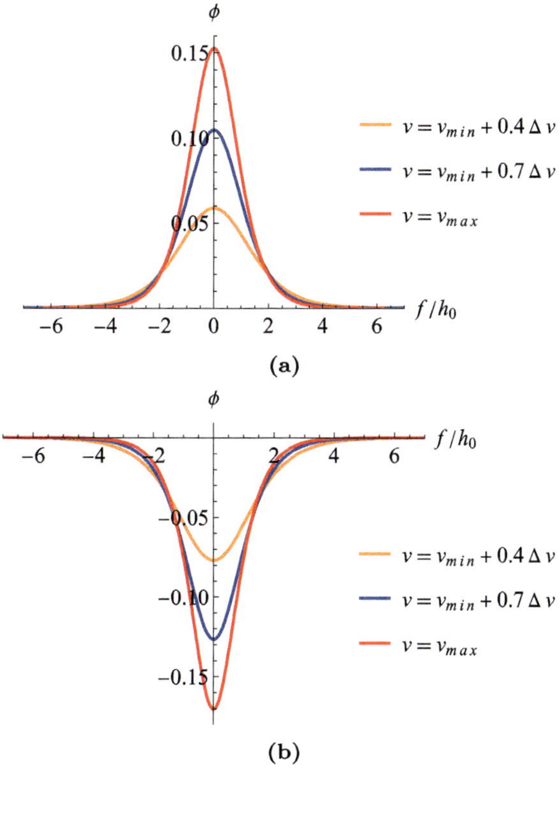

Fig. 4.3 Profiles of solitary waves propagating in tensegrity-like chains equipped with T3D (**a**), and TθD (**b**) units

4.5.2 Overview of Numerical Results Available in the Literature

1D systems We now proceed to review some available literature results on the propagation of compression and rarefaction solitary pulses in 1D, 2D, and 3D arrays of tensegrity units alternating with lumped masses. Comparisons between the numerical results of solitary wave dynamics in tensegrity lattices and the predictions of the analytical models presented in previous sections can be found in Amendola (2023a), de Castro Motta et al. (2024a). We begin by examining the wave dynamics of a 1D mass-spring chain composed of T3 prisms and massive discs, which is impacted at the right end by a compressive disturbance (e.g., an impact with a striker) and subjected to an initial pre-compression force F_0 (Fig. 4.4). We refer the reader to Fraternali et al. (2014), Micheletti et al. (2019) for the geometrical, mechanical and loading properties of the chain examined hereafter, which features millimeter-scale prisms made of Spectra strings and 3D-printable bars composed of the Ti6Al4V titanium alloy. Figure 4.5 shows the propagation of compression solitary strain-waves with compact support in a chain formed by stiffening-type units, while Fig. 4.6 illustrates the intriguing strain-wave dynamics of a chain composed of softening-type units.

The axial strain of the generic unit is denoted by ε and ε_s respectively in Fig. 4.5 and Fig. 4.6. Similarly, the initial strain induced by the pre-compression force F_0 is denoted by ε_0 in Fig. 4.5 and by by ε_{s_0} in Fig. 4.6. The stiffening chain supports compression pulses that span 6–7 units out of a total of 700 units, while the softening chain supports the propagation of a leading rarefaction pulse that spans 10–12 units out of a total of 1400 units. This rarefaction pulse relaxes the initial state of pre-compression in the system and is followed by an oscillatory tail with progressively decreasing amplitude as the wave moves through the system (Fig. 4.6). This generates a metamaterial (or negative-type) response in the chain, as the application of a compression disturbance actually reduces the system's initial state of compression. It is expected that the amplitude of the oscillatory tail would eventually shrink to zero in a sufficiently long chain, whereas one would typically expect an increase in the state of stress due to compressive loading. Such a non-conventional mechanical behavior could be exploited to design novel impact protection devices composed of miniaturized tensegrity lattices with a softening response (Fraternali et al., 2014).

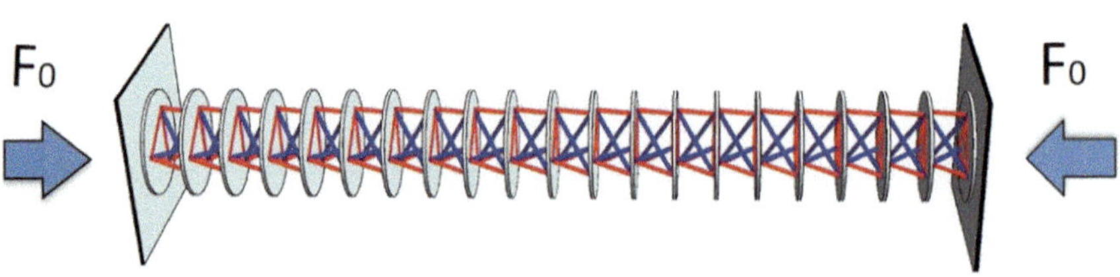

Fig. 4.4 Illustration of a chain formed by alternating tensegrity T3 prisms and discs representing lumped masses, subjected to an initial compression force F_0

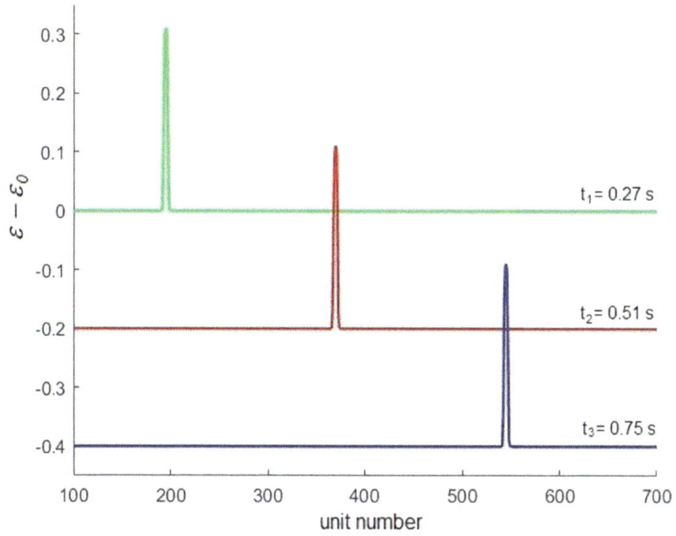

Fig. 4.5 Propagation of a compression solitary strain-wave in a tensegrity mass-spring chain with stiffening response. Reproduced with permission from Micheletti et al. (2019)

Detailed information about the results shown in Figs. 4.5 and 4.6 can be found in Micheletti et al. (2019) and Fraternali et al. (2014), respectively. It is worth noting that these results align with the predictions of the theoretical analysis presented in the previous sections.

2D and 3D systems The wave dynamics of 2D and 3D beams made of arrays of square and cubic tensegrity units has been studied in Micheletti et al. (2019). These units feature centimeter-scale unit cells, Ti6Al4V bars, and Nylon 12 cables. Spherical masses are attached to the nodes, contributing to the mass distribution of the system in conjunction with the masses of the bars. We refer the reader to Table 1 of Micheletti et al. (2019) for the geometric and mechanical properties of the units under consideration. The analyzed beams exhibit a stiffening-type response, which arises from the interpenetration of adjacent cells under compression loading (Micheletti et al., 2019).

Figure 4.7 shows the wave dynamics of a 2D beam formed by a 3×50 grid of square cells. The wave dynamics is generated by imposing an initial velocity v_0 on the central nodes of the left end, while the right end is clamped. The colormaps in the figure illustrate the fractions of energy carried by the members of the tensegrity system at different times. It is important to note that the overall energy of the structure is conserved throughout all the dynamic problems studied here and in the previous sections, as we are not modeling energy dissipation phenomena that could arise from events such as damage or the viscoelastic/viscoplastic response of the members. The results shown in Fig. 4.7 highlight the nucleation of compact compression waves (ccw) ahead of 'thermalized' regions located adjacent to both the impacted (left)

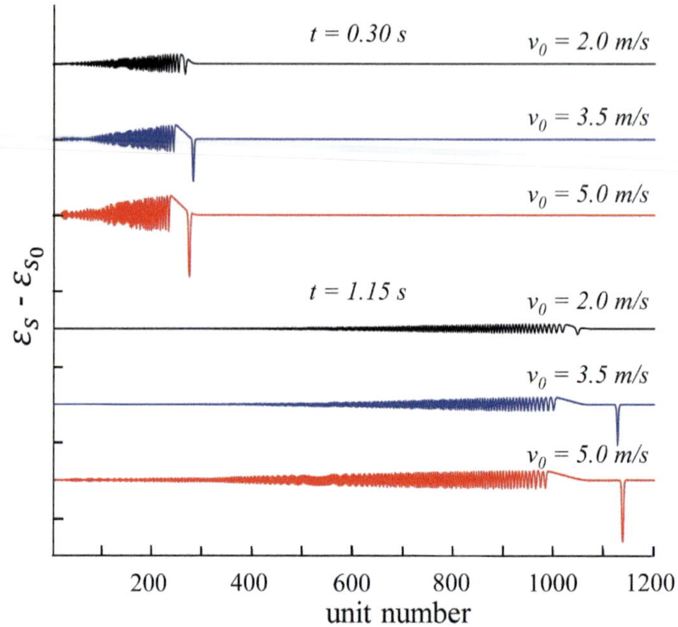

Fig. 4.6 Propagation of a rarefaction solitary strain-pulse followed by an oscillatory tail in a tensegrity mass-spring chain exhibiting a softening-type response, for different values of the impact velocity v_0. Reproduced with permission from Fraternali et al. (2014)

base and the clamped (right) end of the structure. These regions exhibit an agitation motion of the lattice, caused by the impact loading and the wave reflection at the wall, respectively. The traveling ccw spans approximately three columns of unit cells, covering a total of 6 to 9–10 units. This phenomenon generalizes the wave dynamics observed in 1D systems with a stiffening response to two dimensions. It is observed that the 2D ccw experiences limited energy 'leakage' during its propagation through the system, meaning that there is a transfer of energy to adjacent members of the lattice. This leakage is more pronounced after the wave reflects off the wall, amounting to approximately 15%, whereas during the initial propagation from left to right, the energy dispersion at the wave front is much smaller, around 3%.

A similar wave dynamics is observed in a 3D tensegrity beam formed by an array of $2 \times 2 \times 30$ cubic unit cells, which is impacted at the left base and clamped at the right base (Fig. 4.8). Once again, the nucleation of a ccw is seen in front of the impacted area, where thermalization of the lattice occurs, similar to what was noted in the 2D beam analyzed earlier. The total energy fraction $\hat{E}_{ccw}$ carried by the ccw in the 3D beam amounts to approximately 38%, which is significantly higher than the value of $\hat{E}_{ccw}$ observed in the 2D beam, where it reaches around 19% (see Fig. 4.7). The reader is referred to Micheletti et al. (2019) for further details on the dynamic response of 2D and 3D tensegrity beams. The unique wave dynamics observed in these stiffening systems could be leveraged to design novel acoustic lenses with a

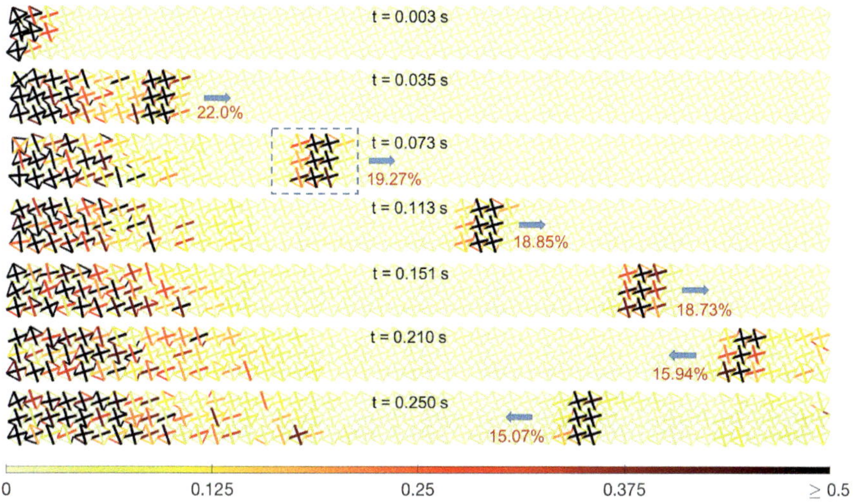

Fig. 4.7 Deformed configurations and energy colormaps at different times for a 2D beam composed of square tensegrity cells, subjected to impact loading at the left end and clamped at the right end. Reproduced with permission from Micheletti et al. (2019)

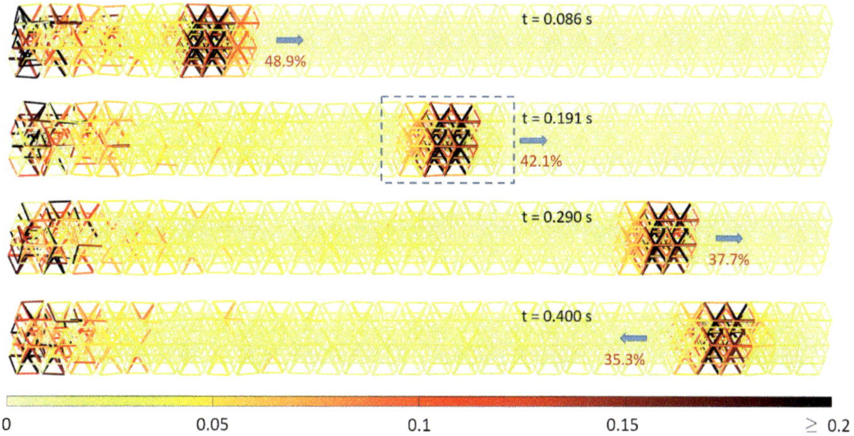

Fig. 4.8 Deformed configurations and energy colormaps at different times for a 3D beam composed of cubic tensegrity cells, subjected to impact loading at the left end and clamped at the right end. Reproduced with permission from Micheletti et al. (2019)

tensegrity architecture, capable of generating compact compression waves in a host medium. Such lenses could be used, for example, to target defects or tumor masses within the host medium (Spadoni & Daraio, 2010; Donahue et al., 2014).

4.6 Concluding Remarks

This Chapter has reviewed the use of tensegrity-like potentials and their role in predicting the existence and properties of solitary waves traveling along tensegrity mass-spring chains. Using a quasi-continuum model and an analysis *à la* Weierstrass of the solutions to the equations of motion (Saccomandi, 2004; Destrade et al., 2007), we have obtained analytic results that confirm previous numerical observations of the existence of solitary pulses in mass-spring systems with tensegrity architecture.

In particular, we have analytically demonstrated that stiffening-type tensegrity mass-spring chains support the propagation of compression solitary pulses (Fraternali et al., 2012; Micheletti et al., 2019; Amendola, 2023a; de Castro Motta et al., 2024b), while initially pre-compressed tensegrity mass-spring systems with a softening-type response support the propagation of rarefaction solitary pulses (Fraternali et al., 2014; de Castro Motta et al., 2024c). We have obtained analytical expressions for wave speeds and pulse shapes in these systems.

Since tensegrity systems have a tunable nature, the solitary waves that propagate in tensegrity lattices can also be tuned for customized uses, increasing their practical applicability. As previously noted, viable technical applications of these systems include the design of tunable acoustic lenses capable of focusing solitary waves (Spadoni & Daraio, 2010; Donahue et al., 2014; Micheletti et al., 2019), the development of actuators and sensors for nondestructive testing (Ni & Rizzo, 2012; Jalali et al., 2023), and innovative impact protection devices designed to transform compressive disturbances into solitary rarefaction pulses (Fraternali et al., 2014). The future directions of the research presented in this Chapter should focus on the experimental validation of the theoretical predictions, the analysis of system stability, and expanding the study to two- and three-dimensional systems. This will aim to increase the applicability of solitary wave dynamics in tensegrity systems, potentially utilizing rapid prototyping techniques to fabricate physical models (Al Sabouni-Zawadzka et al., 2024). The tunable wave dynamics of tensegrity systems also hold potential for applications in vibration isolation and seismic engineering (Fraternali et al., 2018, 2021; Santos et al., 2022; Amendola et al., 2023b; Placidi et al., 2024), which could form the subject of additional future extensions of the present research.

References

Al Sabouni-Zawadzka, A., Gilewski, W., Nazifi Charandabi, R., & Zawadzki, A. (2024). Stability of tensegrity-inspired structures fabricated through additive manufacturing. *Composite Structures, 345*, 118377.

Amendola, A. (2023a). An analytic study on the properties of solitary waves traveling on tensegrity-like lattices. *International Journal of Non-Linear Mechanics, 148*, 104264.

Amendola, A., Mattei, O., Milton, G. W., & Seppecher, P. (2023b). The obstacle problem in no-tension structures and cable nets. *Proceedings of the Royal Society A, 479*(2269), 20220229.

de Castro Motta, J., Fraternali, F., & Saccomandi, G. (2024a). Rarefaction pulses on tensegrity lattices are just sech^2-solitary (dark) waves. *Meccanica*.

de Castro Motta, J., Garanger, K., & Rimoli, J. J. (2024). Propagation of compression solitary waves on tensegrity-like lattices made of truncated octahedrons. *International Journal of Non-Linear Mechanics, 162*, 104716.
de Castro Motta, J., Placidi, L., Nazifi Charandabi, R., & Fraternali, F. (2024). A Weierstrass approach to the analysis of rarefaction solitary waves in tensegrity mass-spring systems. *Communications in Applied and Industrial Mathematics, 15*(2), 8–16.
Destrade, M., Gaeta, G., & Saccomandi, G. (2007). Weierstrass's criterion and compact solitary waves. *Physical Review E, 75*(4), 047601.
Donahue, C. M., Anzel, P. W. J., Bonanomi, L., Keller, T. A., & Daraio, C. (2014). Experimental realization of a nonlinear acoustic lens with a tunable focus. *Applied Physics Letters, 104*(1).
Fraternali, F., Carpentieri, G., Amendola, A., Skelton, R. E., & Nesterenko, V. F. (2014). Multiscale tunability of solitary wave dynamics in tensegrity metamaterials. *Applied Physics Letters, 105* (20).
Fraternali, F., Senatore, L., & Daraio, C. (2012). Solitary waves on tensegrity lattices. *Journal of the Mechanics and Physics of Solids, 60*(6), 1137–1144.
Fraternali, F., Carpentieri, G., & Amendola, A. (2015). On the mechanical modeling of the extreme softening/stiffening response of axially loaded tensegrity prisms. *Journal of the Mechanics and Physics of Solids, 74*, 136–157.
Fraternali, F., Amendola, A., & Benzoni, G. (2018). Innovative seismic isolation devices based on lattice materials: A review. *Ingegneria Sismica/International Journal of Earthquake Engineering, 35*(4), 93–113.
Fraternali, F., Singh, N., Amendola, A., Benzoni, G., & Milton, G. W. (2021). A biomimetic sliding-stretching approach to seismic isolation. *Nonlinear Dynamics, 106*(4), 3147–3159.
Herbold, E. B., & Nesterenko, V. F. (2013). Propagation of rarefaction pulses in discrete materials with strain-softening behavior. *Physical Review Letters, 110*(14), 144101.
Jalali, H., Misra, R., Dickerson, S. J., & Rizzo, P. (2023). Remote wireless monitoring of localized corrosion using compact solitary waves. *Structural Health Monitoring, 22*(3), 1876–1892.
Micheletti, A., Ruscica, G., & Fraternali, F. (2019). On the compact wave dynamics of tensegrity beams in multiple dimensions. *Nonlinear Dynamics, 98*(4), 2737–2753.
Ni, X., & Rizzo, P. (2012). Use of highly nonlinear solitary waves in nondestructive testing. *Materials Evaluation, 70*(5), 561–569.
Placidi, L., de Castro Motta, J., Nazifi Charandabi, R., & Fraternali, F. (2024). A continuum model for the tensegrity Maxwell chain. *NHM, 19*(2), 597–610.
Rosenau, P. (1986). Dynamics of nonlinear mass-spring chains near the continuum limit. *Physics Letters A, 118*(5), 222–227.
Saccomandi, G. (2004). Elastic rods, weierstrass' theory and special travelling waves solutions with compact support. *International Journal of non-linear Mechanics, 39*(2), 331–339.
Santos, F. A., Caroço, C., Amendola, A., Miniaci, M., & Fraternali, F. (2022). 3D tensegrity braces with superelastic response for seismic control. *International Journal for Multiscale Computational Engineering, 20*(5), 53–64.
Spadoni, A., & Daraio, C. (2010). Generation and control of sound bullets with a nonlinear acoustic lens. *Proceedings of the National Academy of Sciences, 107*(16), 7230–7234.

Chapter 5
Flexible Tensegrity Structures: From Space Applications to 3D Tensegrity Metamaterials

Julian J. Rimoli, Kévin Garanger, and Franco Ruffini

Abstract The concept of flexible tensegrity is discussed in this chapter. Flexible tensegrity structures extend the classical definition of tensegrity to allow the bending and buckling of compression members, resulting in large deformations of these members and off-axial forces. These effects need to be accounted for for highly dynamic applications and large deformations of some tensegrity structures with slender bars. A reduced order model of the bars that accounts for these effects is therefore presented. This model is based on a spring-mass system and allows to capture the elastic buckling behavior of bars in compression accurately and efficiently. Because of its reduced number of degrees of freedom, this model is particularly adapted for large-scale or fast simulations of tensegrity structures, in both dynamic and quasi-static settings. We also discuss the advantage of accounting for the elastic buckling of the bars in the design of tensegrity structures, as a mean to increase their capacity to transfer kinetic energy to elastic energy without structural failure. Some possible applications of flexible tensegrity structures are discussed, including a planetary lander and a 3D tensegrity metamaterial. Both are based on a truncated octahedron tensegrity cell, which has the remarkable property of remaining stable under large deformations despite the buckling of its bars.

5.1 Introduction

In the context of tensegrity structures and as seen thus far in this book, compression members are either rigid or subject to infinitesimal displacements. For the latter case, struts or bars can be modeled as linear elastic springs for the purpose of form-finding

J. J. Rimoli (✉) · K. Garanger · F. Ruffini
Department of Mechanical and Aerospace Engineering, University of California, Irvine, CA, USA
e-mail: jrimoli@uci.edu

K. Garanger
e-mail: kevin.garanger@uci.edu

F. Ruffini
e-mail: f.fraternali@unisa.it

F. Fraternali and J. J. Rimoli (eds.), *Tensegrity Systems*, CISM International Centre for Mechanical Sciences 2, https://doi.org/10.1007/978-3-031-82283-4_5

and static analysis. Conditions that would violate the assumption of infinitesimal displacements of these bars, such as buckling, are therefore considered as failure modes of the structure. With this traditional approach, the load at which buckling occurs thus represents a limit that should not be exceeded during the structure's operation.

In some applications involving highly dynamical events, however, ensuring that a tensegrity structure's bars do not buckle may be impractical or overly conservative. For example, high-velocity impacts, which have been considered as part of the normal operation of various envisioned tensegrity systems, are situations where the no-buckling assumption may not hold anymore. For this reason, there is a need to develop a more general framework to describe the behavior of tensegrity structures where the bars may experience large deformations, including buckling. The term "flexible tensegrity system" is proposed to describe such structures.

As with all problems involving the mathematical modeling of physical systems, a balance must be struck between the accuracy of the model and its ease of use. For problems that involve physical simulations, a primary concern related to the ease of use of the model is its computational complexity. This concern becomes especially relevant for tensegrity systems in a variety of applications. These include, for example, large-scale systems involving many bars, such as tensegrity lattices, systems with relatively stiff members requiring small integration time steps during dynamic simulations, and design and control problems, for which many simulations of a system may be required.

To answer the need for a computationally efficient model of flexible tensegrity systems, we propose a reduced order model of bars designed to capture their buckling behavior. This model, described in Sect. 5.2, is based on a discretized spring system with only a few degrees of freedom. The various parameters of the model are identified analytically to closely match the behavior of a bar under various conditions, including buckling. This model is also validated against other more complex models to ensure its accuracy.

The developed model is then used in two examples to demonstrate its capabilities. These examples also provide an opportunity to review some key computational techniques for the static and dynamic analysis of flexible tensegrity systems. The first one, presented in Sect. 5.3, consists of a tensegrity structure used as a planetary lander. A dynamic simulation of the structure under high-velocity impacts is performed to demonstrate the model's ability to understand the behavior of the structure under such conditions. Integration techniques for the dynamic simulation are also discussed. The second example, presented in Sect. 5.4, involves a tensegrity lattice metamaterial used for energy absorption. We show through numerical examples that this lattice remains stable under large compressive strains, even when the bars are buckling. This analysis is performed using a quasistatic approach that is based on a conjugate gradient descent method that we describe in detail.

Employed Notation

The continuous bar's parameters referred to in this chapter are represented by the following symbols: Young's modulus E, cross-sectional area A, length L, and mass density ρ. As for the reduced order model of the bar, the symbols l, k, and m, where a subscript may be added, represent lengths, stiffnesses, and masses, respectively. Vectors are denoted by boldface letters, such as nodal displacements $\boldsymbol{x}$, nodal forces $\boldsymbol{f}$, and unit vectors representing directions $\boldsymbol{u}$.

5.2 Modeling Flexible Tensegrity Systems

Modeling approaches for classical tensegrity structures have traditionally been motivated by form-finding applications (in the sense of equilibrium determination) and quasistatic problems. The assumptions of pure axial forces experienced by the bars and strings, which is implicitly based on the existence of perfect joints at their ends, implies the absence of bending, shear, or torsion in the tensegrity members. Consequently, the failure modes that are normally accounted for in tensegrity analysis techniques for structural members are limited to the yielding of the strings and the buckling of the bars. In particular, the Euler's buckling load of the bars is considered as an upper limit of the force they can withstand before the structure is assumed to fail.

While these assumptions are sufficient for the analysis of quasi-static problems around the original equilibrium configuration and for the form-finding of tensegrity structures, they break down when the structure is subjected to either large deformations or highly dynamic loads, as bars may depart from the infinitesimal deformation assumption and imperfect joints may result in off-axis loads. To address this concern, methods based on the finite element method for the modeling of off-axis loads in bars have been proposed, via a fully Lagrangian (Kebiche et al., 1999; Ben Kahla & Kebiche, 2000) or co-rotational (Zhang et al., 2015) formulation. These methods are able to capture complex behaviors of tensegrity structures, but are computationally expensive and far from the simplicity of traditional tensegrity models. These caveats can make them impractical for the modeling tensegrity structures within the context of robotics and control frameworks, where faster-than-real-time simulations are the norm.

5.2.1 *A Reduced Order Model to Account for Bar Buckling*

An alternative approach to capture the dynamic and large-deformation behavior of the bars in tensegrity structures consists of using a reduced order model with few degrees of freedom (Rimoli, 2018), described thereafter. This model is based on the discretization of the bar into a series of masses and springs, whose parameters are determined to match properties of interest of the equivalent continuum system.

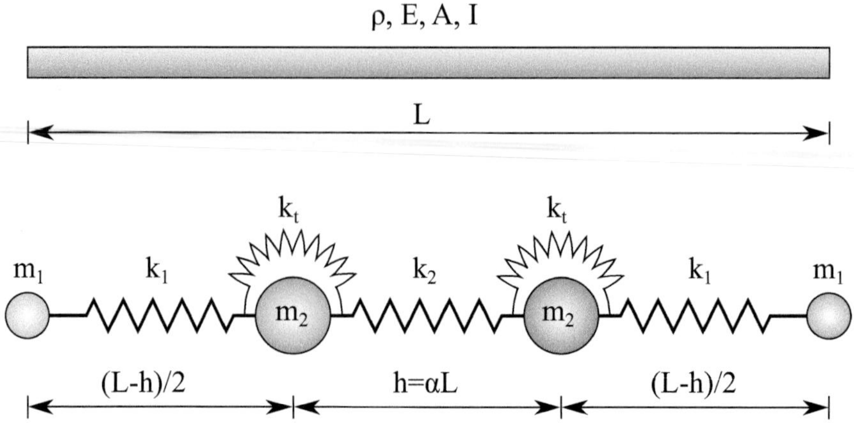

Fig. 5.1 Schematic of the discretization scheme. The continuum bar shown at the top of the figure is replaced by the discrete system shown at the bottom

Model description. Let us consider a bar of length L and constant cross-section of area A. For simplicity, its cross-section is assumed to be circular (as in a rod or a pipe), such that its second moment of area along any axis is constant and denoted by I. The case of a non-circular cross-section can be treated in a similar manner by distinguishing the principal axes of the cross-section. The material of the bar is considered homogeneous and isotropic with Young's modulus E and density ρ. The bar is discretized into a set of four point masses, three linear axial springs, two angular springs, as shown at the bottom of Fig. 5.1. Since the bar has constant cross-section, it is natural to assume that the axial springs at each end of the bar are identical, as well as the two angular springs. The same can be said of the two point masses at the ends of the bars, as well as the two point masses in the middle of the bar. The parameters of the model are therefore:

- l_1 and k_1: length and stiffness of the end axial springs;
- l_2 and k_2: length and stiffness of the middle axial spring;
- k_t: stiffness of the angular springs;
- m_1 and m_2: mass of the point masses at the ends and in the middle of the bar, respectively.

The dynamical state of a bar is then described by the position and velocity of all four masses, resulting in 12 degrees of freedom.

Springs Given ρ, E, A, I, and L, the choice of the reduced order model parameters is performed with the goal of matching some of the key properties of the continuum bar it is based on. The first obvious requirement is that the length of the bar is preserved, resulting in $l_1 = L(1-\alpha)/2$ with $0 < \alpha < 1$ defined such that $l_2 = \alpha L$. α is introduced here as a parameter that can be used to tune certain properties of the model, as seen later in this section.

Constants k_1 and k_2 are then obtained by the standard formula for the axial stiffness of a linear elastic bar, that is

$$k_1 = \frac{2EA}{L(1-\alpha)} \tag{5.1}$$

$$k_2 = \frac{EA}{\alpha L}, \tag{5.2}$$

which incidentally also ensures that the total axial stiffness of the bar is preserved.

The stiffness of the angular springs k_t is chosen such that the discrete bar buckles at the same load as the continuum one. The value to achieve this result is obtained by assuming that a symmetric buckling load occurs under an axial force P, as represented in Fig. 5.2.

Using the generalized coordinates d_1, d_2, and θ to represent the state of the discrete bar as shown in Fig. 5.2, the potential energy of the system is given by

$$\Pi = k_1\left(d_1 - \frac{L(1-\alpha)}{2}\right)^2 + \frac{k_2}{2}(d_2 - \alpha L)^2 + k_t\theta^2. \tag{5.3}$$

Since the displacement induced by P can be expressed as $\Delta L = L - 2d_1\cos(\theta) - d_2$, the equilibrium equations are obtained by equating the derivative of the potential energy and of the work done by P with respect to the generalized coordinates:

$$2k_1\left(d_1 - \frac{L(1-\alpha)}{2}\right) + 2P\cos(\theta) = 0 \tag{5.4}$$

$$k_2\,(d_2 - \alpha L) + P = 0 \tag{5.5}$$

$$2k_t\theta - 2Pd_1\sin(\theta) = 0. \tag{5.6}$$

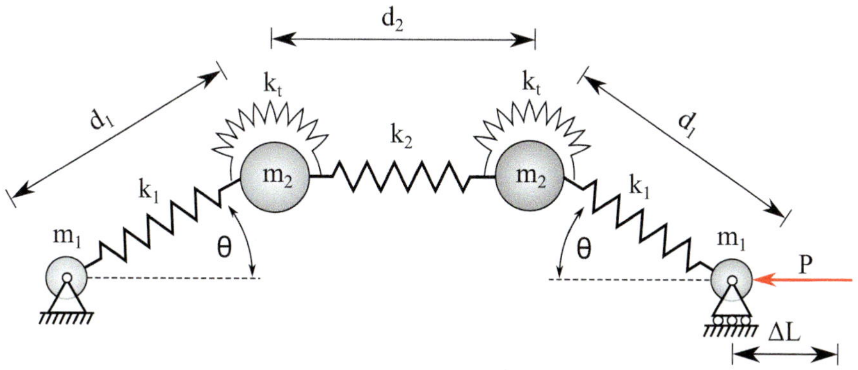

Fig. 5.2 Schematics of buckling configuration for the discretization scheme. A symmetric buckling mode is assumed

After eliminating d_1 from Eq. (5.6) with (5.4) and linearizing the resulting equation around $\theta = 0$, the quadratic polynomial in P

$$2k_t\theta - P\theta L + 2P^2\frac{\theta}{k_1} + P\alpha L = 0 \tag{5.7}$$

is obtained.

Solving Eq. (5.7) for $\theta \neq 0$ yields the critical load P_c at which buckling occurs:

$$P_c = \frac{1}{4}k_1(1-\alpha)L - \frac{1}{4}\sqrt{k_1^2(1-\alpha)^2 L^2 - 16k_1k_t}. \tag{5.8}$$

An interesting observation is that the critical load is independent of the stiffness of the middle spring k_2, although the displacement at which buckling occurs is not. In addition, to ensure that Eq. (5.5) has a solution with $0 < d_2 < \alpha L$, the following condition must be satisfied:

$$k_2 > \frac{P_c}{\alpha L}, \tag{5.9}$$

which is always true since $k_2 = \frac{EA}{\alpha L}$ and $\frac{P_c}{\alpha L} > \frac{1}{4}k_1\frac{1-\alpha}{\alpha} = \frac{1}{2}\frac{EA}{\alpha L}$.

Since the Euler's buckling load of the continuum bar is given by

$$P_c = \pi^2 EI/L^2, \tag{5.10}$$

substituting the value of k_1 from Eqs. (5.1) in (5.8) and equating it to Eq. (5.10) yields the value of k_t:

$$k_t = \frac{1-\alpha}{2}\frac{\pi^2 EI}{L}\left(1 - \frac{I\pi^2}{AL^2}\right) \tag{5.11}$$

Note that for slender bars, $I \ll AL^2$, and the value of k_t simplifies to

$$k_t \approx \frac{1-\alpha}{2}\frac{\pi^2 EI}{L}, \tag{5.12}$$

which is the value that would have been found by neglecting the axial stiffness of the bars, an assumption that is, in fact, made when deriving the Euler's buckling load of a continuum bar.

Masses. The properties determined so far include the ones relevant for quasi-static problems, that is the stiffness of the axial springs k_1 and k_2 and the torsional stiffness k_t. On the other hand, the dynamic properties of the reduced order model are also affected by the value of the masses. These masses are determined by equating the kinetic energy of both bars when subjected to a linear velocity field given by:

$$u(s) = c_1 s + c_2, \quad v(s) = c_3 s + c_4, \quad w(s) = c_5 s + c_6, \tag{5.13}$$

where s is a coordinate along the bar. This velocity field can represent any combination of a rigid body motion with axial deformation of the bars, where c_1, c_3, and c_5 are the components of the velocity field, and c_2, c_4, and c_6 represent the expansion and rotation of the bar.

The kinetic energy of the continuum bar under this velocity field is given by

$$K_{\text{cont}} = \frac{1}{2} \int_{-L/2}^{L/2} \rho A \left(u^2 + v^2 + w^2\right) \mathrm{d}s, \tag{5.14}$$

$$= \frac{1}{24} \rho A L^3 \left(c_1^2 + c_3^2 + c_5^2\right) + \frac{1}{2} \rho A L \left(c_2^2 + c_4^2 + c_6^2\right), \tag{5.15}$$

while the one of the discrete bar is

$$K_{\text{disc}} = \left(\frac{1}{4} m_1 + \frac{1}{4} m_2 \alpha^2\right) L^2 (c_1^2 + c_3^2 + c_5^2) + \frac{1}{2} (m_1 + m_2)(c_2^2 + c_4^2 + c_6^2). \tag{5.16}$$

Ensuring $K_{\text{cont}} = K_{\text{disc}}$ for any velocity field yields the equations

$$m_1 + m_2 = \frac{1}{2} \rho A L, \tag{5.17}$$

$$\frac{1}{4} m_1 + \frac{1}{4} m_2 \alpha^2 = \frac{1}{24} \rho A L^3. \tag{5.18}$$

It is worth noting that Eqs. (5.17) and (5.18) are equivalent to enforcing that the mass and moment of inertia of the discrete bar are equal to the ones of the continuum bar. Solving these equations finally yields the values of m_1 and m_2:

$$m_1 = \frac{1}{6} \rho A L \frac{1 - 3\alpha^2}{1 - \alpha^2}, \tag{5.19}$$

$$m_2 = \frac{1}{3} \rho A L \frac{1}{1 - \alpha^2}. \tag{5.20}$$

It can be seen that for m_1 to be well defined (i.e. positive), the condition $\alpha < 1/\sqrt{3}$ must be satisfied.

Choosing. $\boldsymbol{\alpha}$ Up to this point, the chosen parameters for the discrete bar ensure that it has the same stiffness, mass, moment of inertia, and critical buckling load as the continuum bar, under the condition that $0 < \alpha < 1/\sqrt{3}$. Choosing the specific value of this parameter α can be done to match other specific properties of the continuum bar. For example, it may be desirable to match the frequency response of the continuum bar as closely as possible. To do so, the natural frequencies of the discrete system around the stress-free configuration can be computed as a function of α and compared to the natural frequencies of the continuum bar. In this study, the bending vibration modes are assumed to be uncoupled to the axial modes in order to

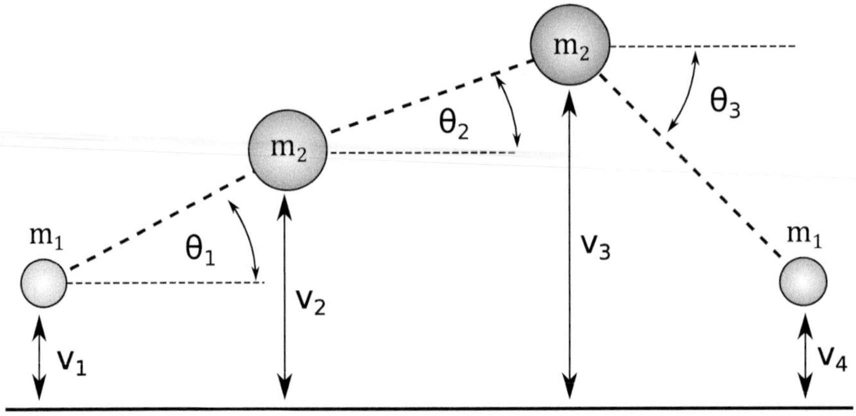

Fig. 5.3 Schematics of the discrete bar's configuration for the computation of the bending mode's natural frequencies

be studied independently. It is the bending modes that will be used to determine α, since the axial modes are only affected by the total axial stiffness of both bar models, which is already matched by construction.

Let the state of the system be described with respect to a reference line by the vertical displacements v_1, v_2, v_3, and v_4, resulting in the angles θ_1, θ_2, and θ_3 as shown in Fig. 5.3, where no axial displacement is considered. With this parameterization, the bending strain energy of the bar is given by

$$U = \frac{1}{2}k_t(\theta_2 - \theta_1)^2 + \frac{1}{2}k_t(\theta_3 - \theta_2)^2. \tag{5.21}$$

Linearizing Eq. (5.21) around the stress-free configuration and substituting k_t from Eq. (5.12) yields

$$\begin{aligned} U =& \frac{\pi^2 EI(2\,\alpha v_1 - \alpha v_2 - \alpha v_3 - v_2 + v_3)^2}{4L^3\,(1-\alpha)\,\alpha^2} \\ &+ \frac{\pi^2 EI(\alpha v_2 + \alpha v_3 - 2\alpha v_4 - v_2 + v_3)^2}{4L^3\,(1-\alpha)\,\alpha^2}. \end{aligned} \tag{5.22}$$

The resulting stiffness matrix of the discrete system is computed by taking the second derivative of the strain energy with respect to the displacements, resulting in

$$\boldsymbol{K}_{\text{bar}} = \begin{bmatrix} -\frac{2\pi^2 EI}{L^3(\alpha-1)} & \frac{\pi^2 EI(\alpha+1)}{L^3(\alpha-1)\alpha} & \frac{\pi^2 EI}{L^3\alpha} & 0 \\ \frac{\pi^2 EI(\alpha+1)}{L^3(\alpha-1)\alpha} & -\frac{\pi^2 EI(\alpha^2+1)}{L^3(\alpha-1)\alpha^2} & -\frac{\pi^2 EI(\alpha+1)}{L^3\alpha^2} & \frac{\pi^2 EI}{L^3\alpha} \\ \frac{\pi^2 EI}{L^3\alpha} & -\frac{\pi^2 EI(\alpha+1)}{L^3\alpha^2} & -\frac{\pi^2 EI(\alpha^2+1)}{L^3(\alpha-1)\alpha^2} & \frac{\pi^2 EI(\alpha+1)}{L^3(\alpha-1)\alpha} \\ 0 & \frac{\pi^2 EI}{L^3\alpha} & \frac{\pi^2 EI(\alpha+1)}{L^3(\alpha-1)\alpha} & -\frac{2\pi^2 EI}{L^3(\alpha-1)} \end{bmatrix}. \tag{5.23}$$

The system is also lumped with the mass matrix $\boldsymbol{M}_{\text{bar}}$, which is given by

$$\boldsymbol{M}_{\text{bar}} = \begin{bmatrix} m_1 & 0 & 0 & 0 \\ 0 & m_2 & 0 & 0 \\ 0 & 0 & m_2 & 0 \\ 0 & 0 & 0 & m_1 \end{bmatrix}. \tag{5.24}$$

The boundary conditions considered here in the context of tensegrity structures are that the ends of the bar are pinned. This condition results in $v_1 = v_4 = 0$ and the first and last rows and columns of the stiffness and mass matrices are removed before the natural frequencies are computed by solving the characteristic equation

$$\det(\boldsymbol{K}_{\text{bar}} - \omega^2 \boldsymbol{M}_{\text{bar}}) = 0. \tag{5.25}$$

The two solutions of Eq. (5.25) are given by

$$\omega_{1,\text{disc}}^2 = 6\,\frac{\pi^2 EJ\,(\alpha+1)}{AL^4\rho}, \qquad \omega_{2,\text{disc}}^2 = 6\,\frac{\pi^2 EJ\,(\alpha+1)}{AL^4\alpha^2\rho}. \tag{5.26}$$

Since α is between 0 and $1/\sqrt{3}$, $\omega_{1,\text{disc}}$ is always lower than $\omega_{2,\text{disc}}$ and is therefore indeed the first natural frequency of the system, while $\omega_{2,\text{disc}}$ is the second one.

As for the continuum bar, its first two natural frequencies when simply supported are given by

$$\omega_{1,\text{cont}}^2 = \frac{\pi^4 EJ}{AL^4\rho}, \quad \omega_{2,\text{cont}}^2 = 4\,\frac{\pi^4 EJ}{AL^4\rho}. \tag{5.27}$$

To assess the accuracy of the discrete model with regard to its natural frequencies, the relative errors of the first and second natural frequencies are computed as

$$\epsilon_1 = \frac{\omega_{1,\text{disc}} - \omega_{1,\text{cont}}}{\omega_{1,\text{cont}}} = \sqrt{\frac{6(\alpha+1)}{\pi^2}} - 1, \tag{5.28}$$

$$\epsilon_2 = \frac{\omega_{2,\text{disc}} - \omega_{2,\text{cont}}}{\omega_{2,\text{cont}}} = \sqrt{\frac{3(\alpha+1)}{2\pi^2\alpha^2}} - 1. \tag{5.29}$$

These errors as a function of α are plotted in Fig. 5.4. It can be seen that for all valid values of α, ϵ_1 is negative, which means that the second natural frequency of the discrete model is always lower than that of the continuum model. Regarding the second natural frequency, the plot of ϵ_2 shows that it is much more sensitive to the choice of α than ϵ_1, and that it is possible to select α such that ϵ_2 is equal to zero. For results presented in the rest of these notes, α is selected such that both errors are equal, which is achieved by setting $\alpha = \frac{1}{2}$ and results in $\epsilon_1 = \epsilon_2 = \frac{3}{\pi} - 1 = -0.045$.

With this choice of α, the natural frequencies of the discrete model underestimates those of the continuum model by 4.5 %, and we also have $\alpha_2 = 2\alpha_1$, which is also the case of the continuum bar. For comparison, a finite element model of the bar with

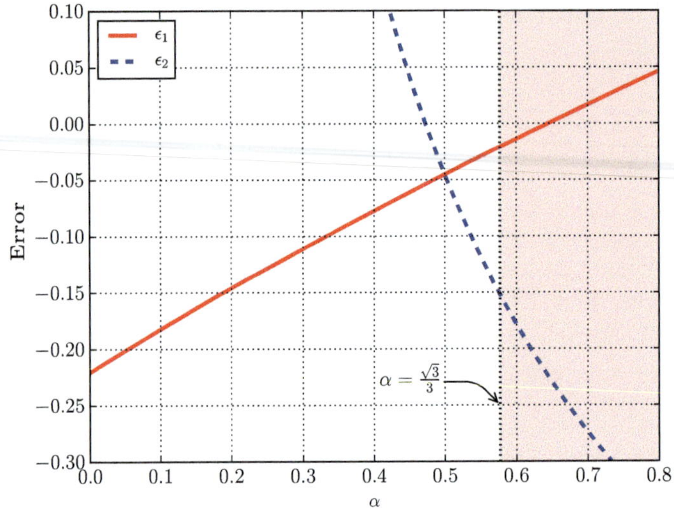

Fig. 5.4 Relative error of the first and second natural frequencies of the discrete model with respect to the continuum model

the same number of degrees of freedom would have a relative error of 18.9 % for the first natural frequency and 9.3 % for the second natural frequency, while resulting in a ratio between these frequencies of 2.28 instead of 2. Obtaining a ratio of 2 between the first and second natural frequencies is important for the dynamical analysis of the tensegrity structure, as it ensures that combines modes of vibrations at these two natural frequencies result in a motion that is periodic in $\frac{2\pi}{\omega_1}$, which would not especially be the case otherwise.

5.2.2 *Elastic Buckling Regime*

Since the introduced reduced order model is meant to characterize the behavior of bars under elastic buckling, it is important to verify that the design parameters of the bars chosen for a structure indeed can result in buckling before any plastic deformation occurs. A simple empirical criterion to determine whether a bar can buckle elastically is given by the Johnson's parabolic formula ((Johnson et al., 1905)). This formula is obtained by fitting a parabola such that it is tangent to the curve that relates the stress at buckling with the slenderness of a column, as obtained from Euler's buckling theory. Defining the slenderness ratio of the examined bar as

$$\kappa = \frac{L\sqrt{A}}{\sqrt{EJ}} \tag{5.30}$$

and using the Euler's buckling load given by (5.10), the critical stress at buckling is given by

$$\sigma_{\text{cr}}^{\text{E}}(\kappa) = \frac{\pi^2 E}{\kappa^2}. \tag{5.31}$$

Defining a parabolic function that is tangent to the curve of $\sigma_{\text{cr}}^{\text{E}}$ and that reaches a maximum of σ_{y} (the material's yielding stress) at $\kappa = 0$ results in the following equation:

$$\sigma_{\text{cr}}^{\text{J}}(\kappa) = \sigma_{\text{y}} \frac{1}{E}\left(\frac{\sigma_{\text{y}}}{2\pi^2}\right)^2 \left(\frac{l}{\kappa}\right)^2, \tag{5.32}$$

with the intersection of the two curves occurring at $\kappa = \sqrt{\frac{2\pi^2 E}{\sigma_{\text{y}}}}$ and $\sigma_{\text{cr}}^{\text{E}} = \sigma_{\text{cr}}^{\text{J}} = \frac{\sigma_{\text{y}}}{2}$. Combining Euler's and Johnson's formulas, the buckling stress of a bar at buckling is given by

$$\sigma_{\text{cr}} = \begin{cases} \sigma_{\text{cr}}^{\text{J}} & \text{if } \kappa < \sqrt{\frac{2\pi^2 E}{\sigma_{\text{y}}}}, \\ \sigma_{\text{cr}}^{\text{E}} & \text{otherwise.} \end{cases} \tag{5.33}$$

The resulting plot of σ_{cr} as a function of κ is shown in Fig. 5.5. The region defined for slenderness values less than the point of tangency represents the plastic regime, while the region defined for slenderness values greater than the point of tangency represents the elastic regime.

Based on the criterion introduced by the Euler's and Johnson's formulas, and in the context of flexible tensegrity structures, the slenderness ratio of the bars should be chosen high enough to ensure that buckling occurs in the elastic regime.

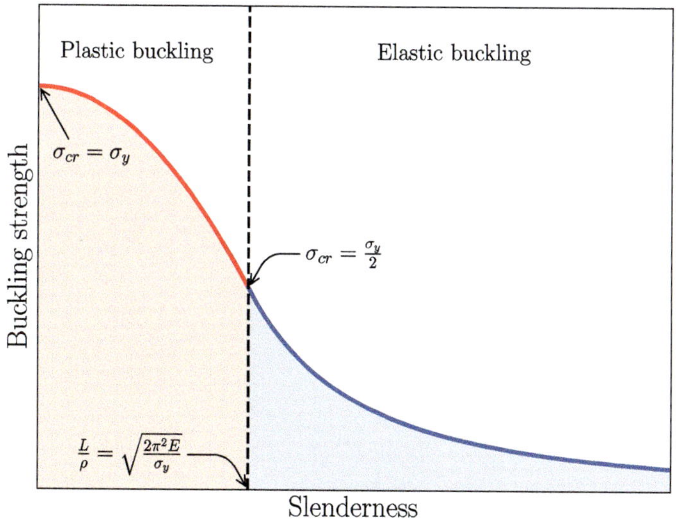

Fig. 5.5 Critical stress at buckling as a function of the slenderness ratio, obtained from Euler's and Johnson's buckling formulas

5.2.3 Numerical Implementation

After having introduced a reduced order model of compressive members for the analysis of flexible tensegrity structures, we now provide some details on the numerical implementation of this model for static and dynamic simulations.

Implementation details. With the goal of implementing the reduced order model of the bar into tensegrity structure simulations, efficient methods of computation of a bar's energy and forces in a given state are needed. Let us assume that the coordinates of a bar are given by the positions of its four masses in an euclidean coordinate system, denoted by $\boldsymbol{x}_1$, $\boldsymbol{x}_2$, $\boldsymbol{x}_3$, and $\boldsymbol{x}_4$, as shown in Fig. 5.6.

By defining the vectors $\boldsymbol{t}_1 = \boldsymbol{x}_2 - \boldsymbol{x}_1$, $\boldsymbol{t}_2 = \boldsymbol{x}_3 - \boldsymbol{x}_2$, and $\boldsymbol{t}_3 = \boldsymbol{x}_4 - \boldsymbol{x}_3$, and their corresponding unit norm vectors $\boldsymbol{u}_1 = \boldsymbol{t}_1/\|\boldsymbol{t}_1\|$, $\boldsymbol{u}_2 = \boldsymbol{t}_2/\|\boldsymbol{t}_2\|$, and $\boldsymbol{u}_3 = \boldsymbol{t}_3/\|\boldsymbol{t}_3\|$, The strain energy of the bar is given by

$$U = \frac{1}{2}\left[k_1\left(\Delta L_1^2 + \Delta L_3^2\right) + k_2 \Delta L_2^2 + k_t\left(\Delta\theta_1^2 + \Delta\theta_2^2\right)\right], \tag{5.34}$$

where

$$\Delta L_1 = \|\boldsymbol{t}_1\| - \frac{1-\alpha}{2}L, \tag{5.35}$$

$$\Delta L_2 = \|\boldsymbol{t}_2\| - \alpha L, \tag{5.36}$$

$$\Delta L_3 = \|\boldsymbol{t}_3\| - \frac{1-\alpha}{2}L, \tag{5.37}$$

$$\Delta\theta_1 = \theta_2 - \theta_1 = \arccos\left(\boldsymbol{u}_1 \cdot \boldsymbol{u}_2\right), \tag{5.38}$$

$$\Delta\theta_2 = \theta_3 - \theta_2 = \arccos\left(\boldsymbol{u}_2 \cdot \boldsymbol{u}_3\right). \tag{5.39}$$

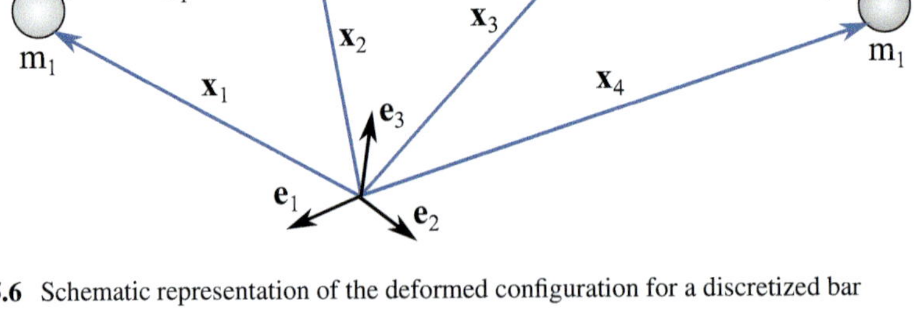

Fig. 5.6 Schematic representation of the deformed configuration for a discretized bar

The elastic forces applied at the masses and resulting from the deformation of the bar are given by

$$\boldsymbol{f}_i = -\frac{\partial U}{\partial \boldsymbol{x}_i} \quad \forall i = 1, 2, 3, 4. \tag{5.40}$$

In the case of $i = 1$, this force vector is given by

$$\begin{aligned} \boldsymbol{f}_1 = & -k_1 \frac{\Delta L_1}{\|\boldsymbol{x}_1 - \boldsymbol{x}_2\|} (\boldsymbol{x}_1 - \boldsymbol{x}_2) \\ & + k_t \frac{\Delta\theta_1}{\sin \Delta\theta_1} \frac{(\boldsymbol{x}_2 - \boldsymbol{x}_1) \times ((\boldsymbol{x}_2 - \boldsymbol{x}_1) \times (\boldsymbol{x}_3 - \boldsymbol{x}_2))}{\|\boldsymbol{x}_2 - \boldsymbol{x}_1\|^3 \|\boldsymbol{x}_3 - \boldsymbol{x}_2\|} \\ = & -k_1 \Delta L_1 \boldsymbol{u}_1 - k_t \frac{\Delta\theta_1}{\sin \Delta\theta_1} \frac{\boldsymbol{u}_1 \times (\boldsymbol{u}_2 \times \boldsymbol{u}_1)}{\|\boldsymbol{t}_1\|}. \end{aligned} \tag{5.41}$$

The ratio of $\Delta\theta_1$ to $\sin \Delta\theta_1$ in (5.41) can be defined by continuity everywhere with a value of 1 for $\Delta\theta_1 = 0$. However, within the context of numerical simulations, the potential division by zero needs to be addressed, especially since it corresponds to the unbuckled state of the bar. To do so, a solution is to use a Taylor expansion $\sin \Delta\theta_1$ around $\Delta\theta_1 = 0$. Such an expansion to the ninth order leads to

$$\boldsymbol{f}_1 = -k_1 \Delta L_1 \boldsymbol{u}_1 + \frac{k_t}{1 - \frac{\Delta\theta_1^2}{6} + \frac{\Delta\theta_1^4}{120} - \frac{\Delta\theta_1^6}{5040} + \frac{\Delta\theta_1^8}{362880}} \frac{1}{\|\boldsymbol{t}_1\|} \boldsymbol{u}_1 \times (\boldsymbol{u}_2 \times \boldsymbol{u}_1). \tag{5.42}$$

A similar expression can be derived for $\boldsymbol{f}_2$, $\boldsymbol{f}_3$, and $\boldsymbol{f}_4$, by also performing a Taylor expansion of $\sin \Delta\theta_2$ around $\Delta\theta_2 = 0$. This process allows for the computation of the forces as demonstrated in Algorithm 5.1.

Numerical example: post-buckling analysis. As a first numerical example of the proposed model, a post-buckling analysis is performed by computing the force-displacement curve of a bar under axial compression. The bar is of length $L_b = 0.58$ m, with a circular tube cross-section of outer diameter 19 m and thickness 1 mm, resulting in $A = 5.62 \times 10^{-5}\,\text{m}^2$ and $I = 2.3 \times 10^{-9}\,\text{m}^4$. The chosen material is titanium, with $E = 91 \times 10^9$ Pa and $\rho = 4480\,\text{kg m}^{-3}$.

This quasi-static analysis is conducted by finding the successive equilibrium states of the bar as the axial displacement is increased from 0 tos 0.01 m. Given the nonlinear nature of the problem and the absence of closed-form solution to the equilibrium equations of the bar in general, each equilibrium state is found numerically by using a variant of the Polak-Ribière variant of the conjugate gradient method. This method is detailed in Algorithm 5.2.

The resulting force-displacement curve is shown in Fig. 5.7. In addition, this curve is compared the one obtained for an elastica model of the bar (Kraus & Rimoli 2023). The pre-buckling linear response of the bar is identical in both models and buckling occurs at the same load, as expected. In the post-buckling regime, the load slightly increases in both models, which is also expected. Although the post-buckling behavior of the reduced order model is not identical to the elastica bar, the difference

```
Input: Nodal positions p
Output: Nodal forces f
function apply_forces(p): begin
    for j ← 0 to 3 do
        t[j] ← p[j + 1] − p[j];
        d[j] ← norm(t[j]);
        u[j] ← t[j]/d[j];
    end
    f ← zeros((4, 3));
    // List of 4 3D vectors
    for j ← 0 to 1 do
        sint ← cross_product(u[j], u[j + 1]);
        cost ← dot_product(u[j], u[j + 1]);
        Δθ ← atan2(cost, norm(sint));
        c ← kt[j]/(1 − Δθ^2/6 + Δθ^4/120 − Δθ^6/5040 + Δθ^8/362880);
        f1 ← c/d[j] · cross_product(u[j], sint);
        f3 ← c/d[j+1] · cross_product(u[j + 1], sint);
        f2 ← −(f1 + f3);
        f[j] ← f[j] + f1;
        f[j + 1] ← f[j + 1] + f2;
        f[j + 2] ← f[j + 2] + f3;
    end
    for j ← 0 to 2 do
        f_ax ← k[j] · (d[j] − d0[j]);
        f[j] ← f[j] + f_ax · u[j];
        f[j + 1] ← f[j + 1] − f_ax · u[j];
    end
end
```

Algorithm 5.1: Reduced order model forces computation

is small with respect to the buckling load. This numerical experiment shows that the reduced order model, which was designed to capture the onset of buckling, also turns out to replicate the post-buckling behavior of the elastica model with good precision. Of particular interest is the fact that the effective response of the system is nonlinear, even thought the discretization is based on linear springs. Clearly, the source of nonlinearity is geometric, as is the case for the continuum bar. In this sense, our discrete model captures the fundamental nature observed by the elastica theory in the postbuckling regime: all sources of nonlinearities are geometric and not material.

5.3 Dynamic Response: The Case for Tensegrity Planetary Lander

Planetary landers based on tensegrity structures have been proposed as a lightweight and robust alternative to traditional systems (SunSpiral et al., 2013; Rimoli, 2016; Garanger et al., 2021), which rely on thrusters to reduce impact velocity and landing

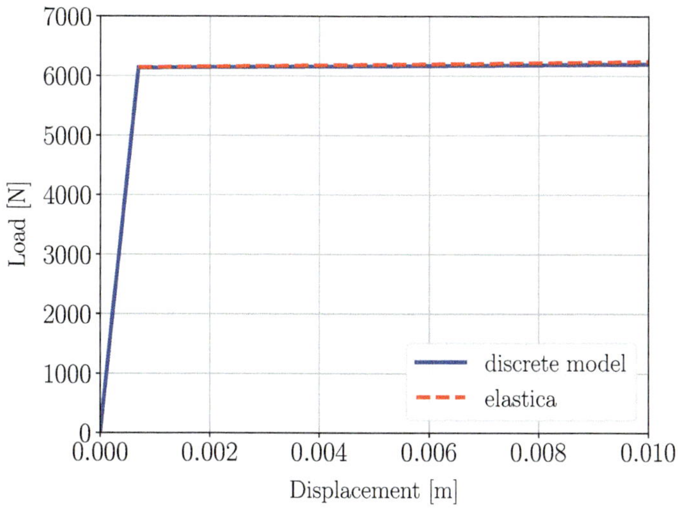

Fig. 5.7 Load-displacement plot for continuum and discrete bars

Input: Nodal positions p, gradient function $DU(\cdot)$, maximum iterations N, tolerance t, small number ε.

$\alpha \leftarrow \beta \leftarrow dr \leftarrow 0$;
$r \leftarrow DU(p)$; $\quad r_n \leftarrow r$; $\quad d \leftarrow r$;
for $(k = 0, \ldots, N)$ **do**
  $x \leftarrow x + \varepsilon \cdot d$;
  $dr \leftarrow \frac{(DU(x)-r)\cdot d}{\varepsilon}$;
  $\alpha \leftarrow \alpha - \frac{r \cdot d}{dr} - \varepsilon$;
  $x \leftarrow x + \alpha \cdot d$;
  $r_n \leftarrow DU(x)$;
  if $(\|r_n\| < t)$ **then**
    update local state;
    return *SUCCESS*;
  end
  $\beta \leftarrow \frac{r_n(r_n - r)}{r^2}$;
  if $(\beta < 0)$ **then**
    $\beta = 0$;
  end
  $d \leftarrow r_n + \beta \cdot d$;
  $r \leftarrow r_n$;
end
return *FAIL*;

Algorithm 5.2: Polak-Ribière algorithm applied for equilibrium finding

gears or airbags to absorb the impact energy. The main advantage of tensegrity landers is that they offer a structure that can be packed in a tight space for transport, then deployed for entry, descent, and landing, upon which the impact energy can be absorbed via the deformation of the structure.

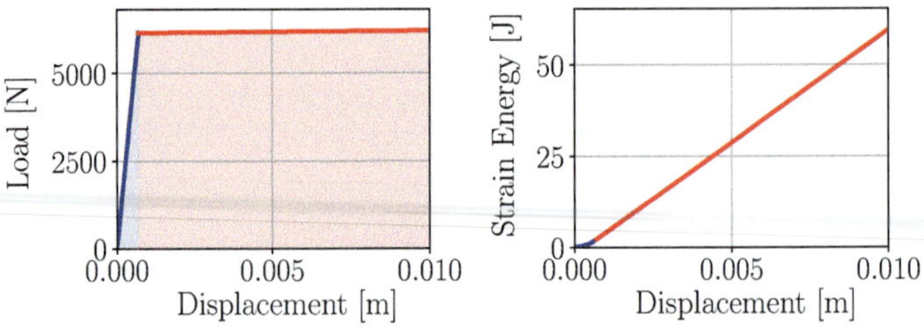

Fig. 5.8 Force-displacement and energy-displacement curves for the bar of Sect. 5.2

Allowing elastic buckling of the bars in a tensegrity structure has the added benefit of increasing the total available elastic energy that can be stored without failure. To illustrate this fact, plots of the force-displacement and energy-displacement curves for the bar of Sect. 5.2 are shown in Fig. 5.8. These plots show that the energy that can be stored in the bar in the post-buckling regime increases significantly in comparison with the one that can be stored up to buckling.

Based on the model of Sect. 5.2, a simulated impact of a tensegrity planetary lander is performed. The lander's geometry is based on a truncated octahedron, which is a tensegrity structure with 12 bars and 36 cables, detailed in (Li et al. 2010). Figure 5.9 shows the lander's geometry, although it's important to highlight that the state represented is without any pre-stressing of cables. The pre-stressing of the cables after the lander is assembled, which is necessary for the structure to be stiff, will invariably affect the shape of the lander in its equilibrium configuration.

As seen in the figure, the lander's structure consists of six square faces that can be grouped into three pairs of parallel faces. These faces can in fact be seen as inscribed in a cube that fully encloses the lander. We refer to these squares as top, bottom, left, right, front, and back.

In the following numerical simulations, the lander's bars are the same as the one described in Sect. 5.2 and therefore have same length $L_b = 0.58$ m. The cables are made of the same material as the bars, have same length $L_b = 0.58$ m, and are 5 m in

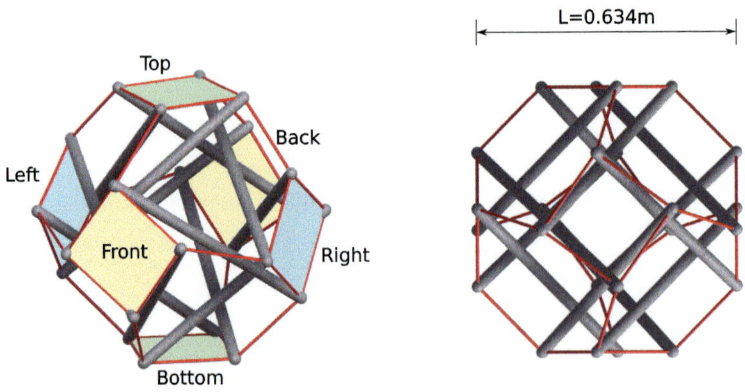

Fig. 5.9 Tensegrity lander based on a truncated regular octahedron

diameter, resulting in a cross-sectional area of 1.96×10^{-5} m^2. The distance between each pair of parallel faces is roughly 0.634 m, which is the side length of the cube that encloses the lander.

Impact tests were carried out at velocities of 3, 6, 9, 12, 15, 18 m s^{-1}, with 3 m s^{-1} representing a typical landing velocity for Mars missions. For each velocity, three different levels of cable pre-stretch were considered, defined by λ, the ratio of the cables stressed length, denoted by L_{ps}, to the cables unstressed length, defined previously as L_c:

$$\lambda = \frac{L_{ps}}{L_c}. \tag{5.43}$$

A value of $\lambda = 1$ corresponds to no pre-stretch, while $\lambda = 0.99$ indicates that the cables have a rest length that is 1 % shorter than their length under tension.

The impact tests are performed with the lander in its equilibrium configuration, which is found by minimizing the strain energy of the structure given in Eq. (5.34) with Algorithm Fig. 5.2. The equilibrium configurations obtained for three different values of λ are shown in Fig. 5.10.

Impacts are then simulated for the various systems at the prescribed velocities. A Newmark time stepping method is used to integrate the dynamics of the system. Contacts between the lander and the ground are modeled using a penalty method for the normal force, where the contact force is proportional to the penetration depth of the lander into the ground. This force is bounded to ensure that the Courant condition is satisfied ((Courant et al., 1967)). Friction forces are not considered during the simulation of the impact.

The evolution of a simulated impact at a velocity of 12 m s^{-1} and a pre-stretch level of 0.999 is shown in Fig. 5.11. The first picture corresponds to a moment right before the impact. A severe deformation of the structure is observed during the impact as time progresses. Until $t = 0.17$ s, the bars of the lander can be seen to remain mostly straight, and most the deformation is due to geometric nonlinearities and stretching of the cables. At $t = 0.21$ s, the bars start to bend, due to a combination of off-axis forces and higher axial loads. Strong bending oscillations can be observed past that point, as that deformation mechanism becomes dominant.

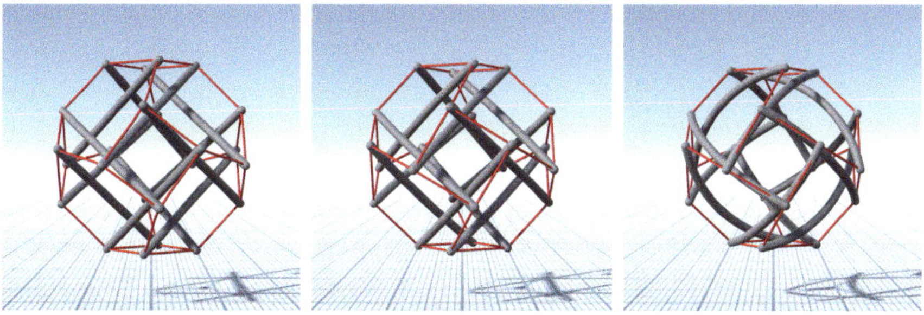

(a) $\lambda = 1$ (b) $\lambda = 0.99$ (c) $\lambda = 0.95$

Fig. 5.10 Tensegrity lander equilibrium for different cable pre-stretch levels

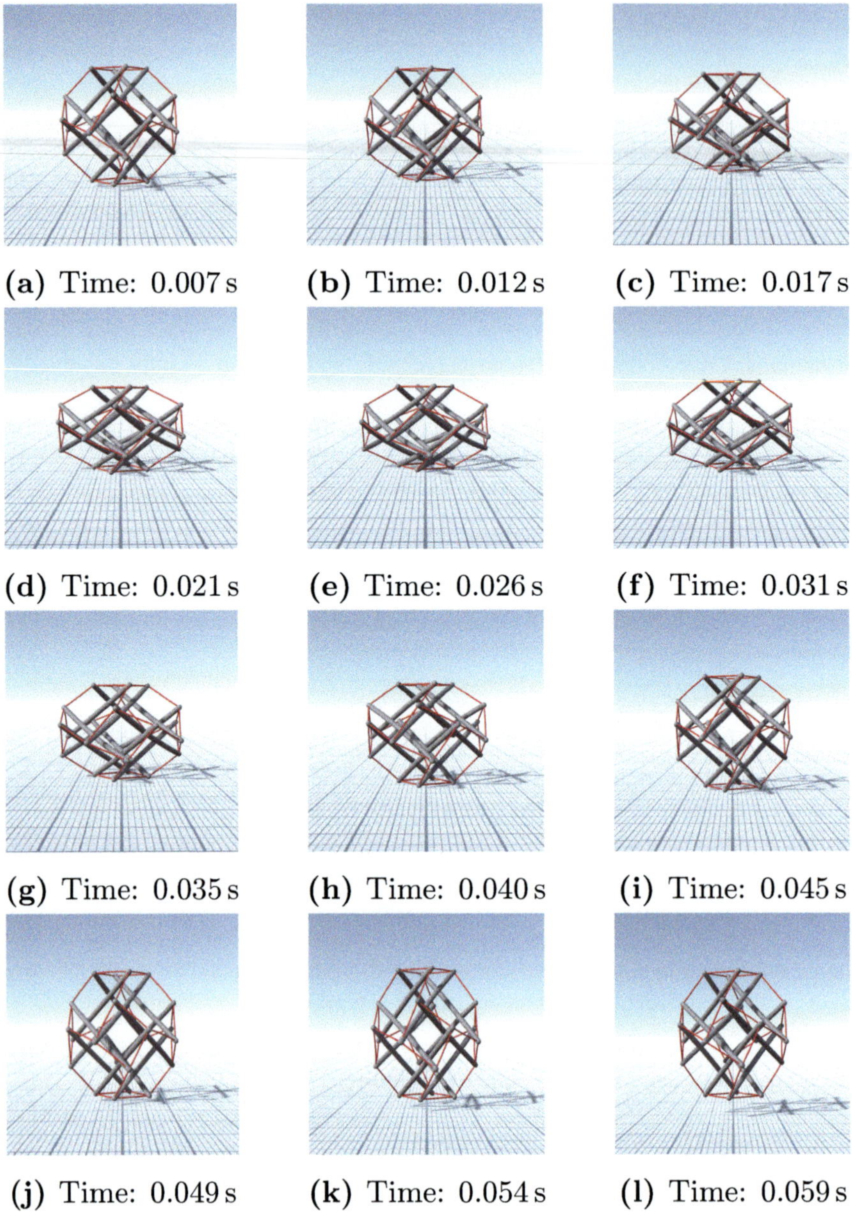

Fig. 5.11 Evolution of the simulated impact test for a tensegrity lander with impact velocity of $12\,\mathrm{m\,s^{-1}}$ and pre-stretch level corresponding to $\lambda = 0.996$

Figure 5.11 offers a qualitative perspective on the simulated impact of the tensegrity lander, while quantitative results are observable from Fig. 5.12. In the latter, the force history of all bars are plotted for various combinations of impact velocities and pre-stretch levels. On the left column, the case with no pre-stretch are depicted, and

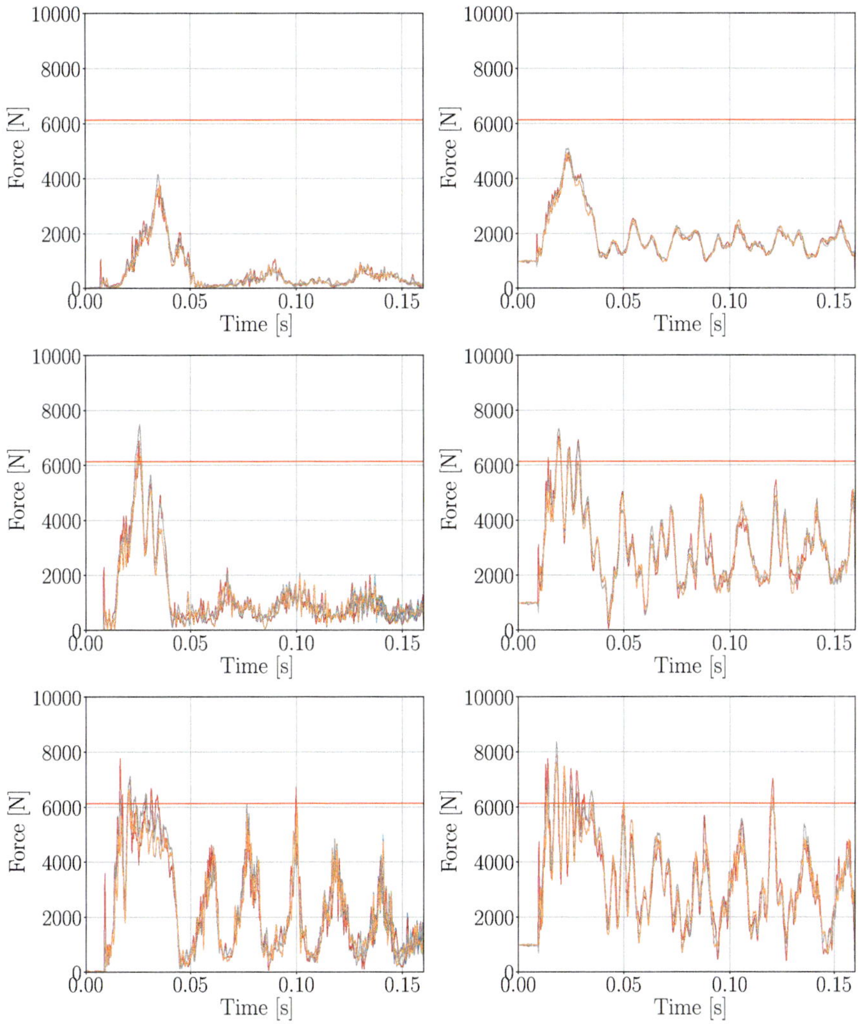

Fig. 5.12 Force history on bars for different impact velocities and cable pre-stress levels. The left column corresponds to $\lambda = 1$ and the right one to $\lambda = 0.994$. The top, middle, and bottom rows correspond to an impact velocity of 6, 12, and 18 $\mathrm{m\,s^{-1}}$, respectively

the right column shows the case of $\lambda = 0.994$. The following impact velocities are displayed, from the top to bottom rows: 6, 12, and 18 $\mathrm{m\,s^{-1}}$. The red line present in all figures indicates the Euler's buckling load of the bars.

In the case with no pre-stretch (left column), for the 6 $\mathrm{m\,s^{-1}}$ impact velocity, the time history shows that the bar loads peak approximately at 0.035 s at a force of 4000 N. The impact lasts about 0.04 s and the initial load as the beginning of the impact, of about 1000 N, is well below the peak impact load. After impact and as the lander bounces back up, macroscopic oscillations can be observed, which correspond

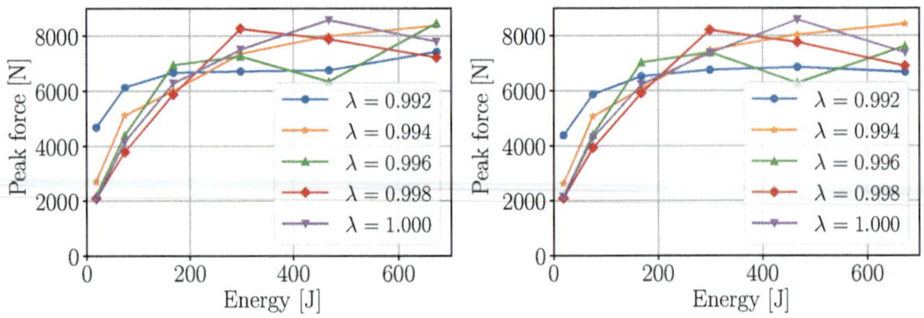

Fig. 5.13 Peak force in bars (left) and cables (right) for different impact energies and cable pre-stretch levels

to an upward worm-like motion of the lander, as well as higher frequency bending oscillations.

For higher velocity impacts, it can be observed that the initial contact load remains below the peak load, which is also achieved earlier. Peak loads tend to saturate as the impact velocity increases, as the bars buckle and reach their maximum load-bearing capacity. Despite the buckling of the bars, it is worth highlighting that the structure does not become unstable. As for the macroscopic oscillations of the lander as it bounces back, a higher impact velocity results in a higher oscillation frequency.

The pre-stressed case exhibit similar features as the unstressed case. However, it can be noted the frequency of the bars' oscillations are reduced when the cables are pre-stressed, which can be explain by the fact they oppose the bar forces and therefore slow down their oscillatory behavior.

An important observation that applies to all cases is that, except for the first load peak that occurs during contact, the force history is relatively similar in the multiple simulated conditions, as the period of the macroscopic oscillations seems to converge to a value of $\approx$0.04 s and dominate the force response of the structure. In addition, the strain energy appears to be evenly distributed among all the bars, resulting in a very efficient energy storage density of the lander.

Plots in Fig. 5.13 show that the peak force in the bars and cables increases with the impact velocity. It can be seen that peak loads for both member types increase as the kinetic energy of the impact does. However, a plateau is eventually reached, as the bars reach their buckling load and cannot sustain higher loads.

5.4 Quasistatic Response: The Case for 3D Tensegrity Metamaterials

5.4.1 Construction of 3D Tensegrity Lattices

In the previous section, it was observed that the truncated octahedron tensegrity structure used for the planetary lander remained stable even after large deformation and buckling of its compression members. This property is a main reason for the capability of the structure to absorb the energy of high-velocity impacts without structural failure. In addition, it was observed that, post-buckling, the energy stored by the structure is mainly concentrated in the bars and evenly spread among them. Based on these findings, we propose to use this truncated octahedron structure to design tensegrity lattices, with the goal of engineering metamaterials capable of absorbing high-energy impacts (Rimoli & Pal, 2017).

Indeed, the failure of material is typically a consequence of localized mechanisms. For truss-like structures, collapse is often initialized by the buckling of a few members, which leads to a redistribution of the loads and the eventual failure of the structure. The capability of the tensegrity structure to distribute the energy among all its members and to remain stable even after buckling of some of its members makes it an interesting candidate for the design of metamaterials that exhibit delocalization, that is metamaterials that do not fail due to the localized failure of its members.

Let us consider the truncated octahedron tensegrity previously introduced, as represented in Fig. 5.9, where the six faces and their denominations are explicitly represented. Although each pair of opposed faces are in parallel planes that incidentally define an enclosing cube, it would not be possible to translate the tensegrity structure along all three directions of space to create a lattice in such a way. Indeed, by looking for example specifically at the top and bottom faces of Fig. 5.9, it can be seen that one of these faces cannot be obtained from the other by a simple translation, as the faces would not coincide. To make these faces coincide, a composition of a translation and of a rotation along the vertical axis would also be required. Stacking a second truncated octahedron cell to the first one by connecting the top face of one of the cells to the bottom face of the other would therefore result in the cells being twisted respectively to each other. A one-dimensional chain of cells assembled this way would be possible, but a two- or three-dimensional tiling of the space would not.

Besides this impossibility of creating a lattice simply by stacking identical cells together, another major issue that comes with this method is that it may result in continuous paths of compression from one end of the structure to the other, as seen on Fig. 5.14. This feature is antithetical to a conventionally accepted requirement for tensegrity structures as initially described by Buckminster Fuller, where tensegrity structures consist of "islands of compression in a sea of tension".

To address these issues, it is possible to create a structure from the truncated octahedron tensegrity through successive reflections that result in a valid lattice unit cell. Specifically, three successive reflections of the original structure must be performed around three of its faces all perpendicular to each other. By choosing, for example, the right, top, and back faces as the ones around which the reflections are performed, as the ones shown in Fig. 5.15, the resulting structure that consists of 8 truncated octahedra verifies translational symmetry around all three canonical axes. That is, the groups of nodes belonging to each opposed face on the boundary of this expanded structure can be obtained from one another by a direct translation along an axis perpendicular to these faces. As a consequence, a three-dimensional cubic lattice can be created from this structure from successive translations of this newly defined supercell along these axes.

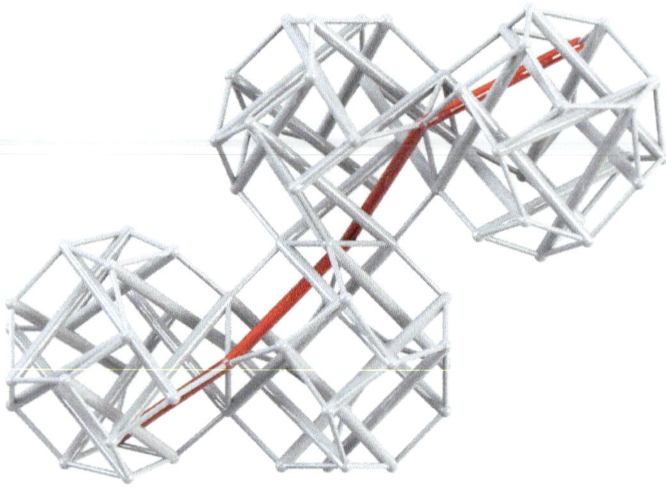

Fig. 5.14 Simply stacking of elementary cells leads to continuous compression paths extending throughout the structure

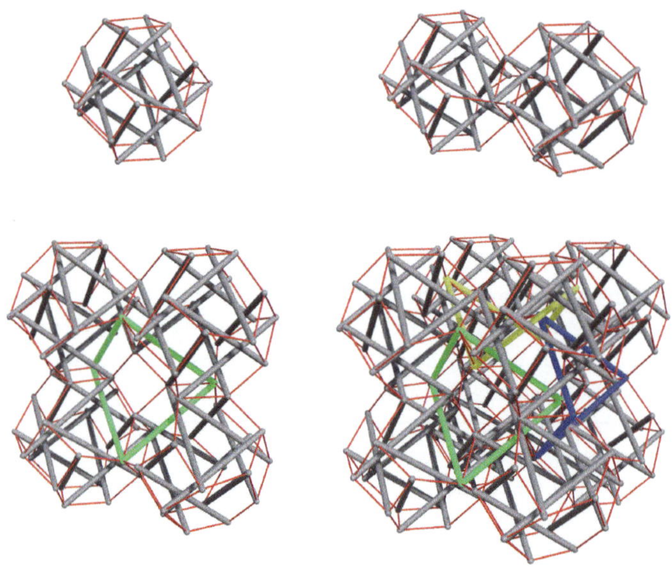

Fig. 5.15 Sequence of reflection operations needed to generate a unit cell compatible with translational symmetries

It is crucial to highlight that an added benefit of this method is that it preserves the fundamental property of tensegrity, according to which compression members must form isolated paths connected together via tension members. Indeed, in Fig. 5.15, it can be seen that compression paths form closed loops and consist of at most four bars. In total, six of these closed loops can be counted in the supercell, while all the other bars are either isolated or forming 2-bar V-shape arrangements that would be part of closed loops, were more cells added to form a three-dimensional lattice. With this construction method, the concept of isolated compression islands in a sea of tension as in Fuller's definition is thus respected.

5.4.2 Delocalization in Tensegrity Lattices

Localized deformation in load-bearing systems leads to system-level instability, i.e., a decrease in load-bearing capacity with increasing deformation, resulting in large deformation increases without further load increase until a new equilibrium is achieved (often after a catastrophic collapse). Conversely, localized deformation can be identified by the presence of system-level instability. On the other hand, a system is stable when increasing deformation leads to increased load-bearing capacity. An increasing load-bearing capacity can be characterized by a positive stiffness on a stress-strain curve along a region of interest.

Simulated response of a 3D tensegrity lattice. The observed stability in the tensegrity lander example is therefore an encouraging sign that tensegrity lattices built from truncated octahedron unit cells may not experience localization even under large deformations. However, the response of the structure made from tiling space with the supercells, themselves constructed from successive reflections of the truncated octahedron structure, necessarily differs from the one of a single cell because of boundary effects and constraints imposed by connecting multiple cells together.

Nonetheless, simulated compression tests of 3D tensegrity lattices have shown that indeed, delocalization leads to a high resistance to failure in these lattices. In simulations, tensegrity lattices of various sizes, volume densities, and material's elastic moduli were subjected to uniaxial compression tests between two parallel plates. These tensegrity lattices were simulated using the Abaqus SIMULIA software and both bars and cables were modeled by discretizing them into multiple beam elements. In all simulations, the structural elements have the same radius and are made of the same isotropic linear elastic material. No pre-stretch was applied to the cables, since the beams elements provide stability to the structure in the absence of pre-stressing. The boundary conditions were set by adding plates at the top and bottom of the lattice, such that the nodes at both ends of the lattice are constrained vertically but free to move horizontally.

In Fig. 5.16, an illustration is shown of a simulated compression test for a $4 \times 4 \times 4$ lattice, that is, a lattice consisting of 64 unit cells and 512 truncated octahedron units.

In particular, compression simulations were performed for five different Young's moduli ranging from 800 to 1200 MPa. The resulting stress-strain curves and energy density curves are shown in Fig. 5.17. It can be seen that in all cases, the structure remains stable over the whole range of deformation (reaching 75 % strain) as the energy density is convex as a function of the strain.

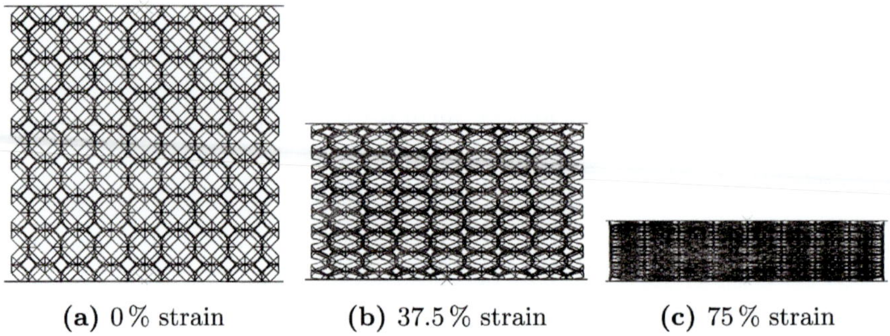

Fig. 5.16 Illustration of the deformation of the 4 × 4 × 4 tensegrity lattice under uniaxial compression

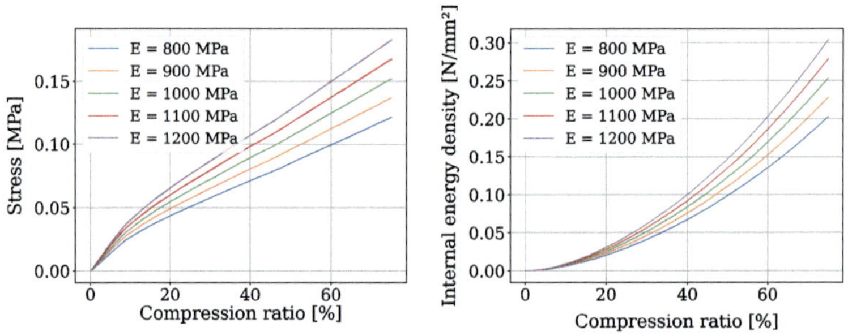

Fig. 5.17 Effect of the elastic modulus on the structure's stress and internal energy density

Effect of material's volume density. Prior studies on lattice metamaterials have shown the existence of a relationship between their mechanical behavior and their volume density ρ_V, defined as the volume fraction of material in the lattice (Deshpande et al. 2001; Gibson & Ashby 1997). For tensegrity lattices of size $n \times m \times l$, it can be shown that this volume density is given by the following equation:

$$\rho_V = \left(\frac{r}{D}\right)^2 \frac{\pi}{8} \left[A + \frac{l + m + n}{lmn} B\right] \tag{5.44}$$

where $A \approx 157.06$ and $B \approx 5.77$ are constants, r is the radius of the bars and cables (assumed equal in this section), and D is the distance between two parallel of the truncated octahedron unit.

We note that for a fixed lattice size, its density is proportional to the square of $\frac{r}{D}$, itself proportional to the slenderness ratio of its members (assuming as well a fixed pre-stretch value). Figure 5.18 shows the effect of the volume density on the stress-strain curves and energy density curves of the tensegrity lattice. It can be seen that the energy density curves are convex for all volume densities, indicating that the structure remains stable over the whole range of deformation.

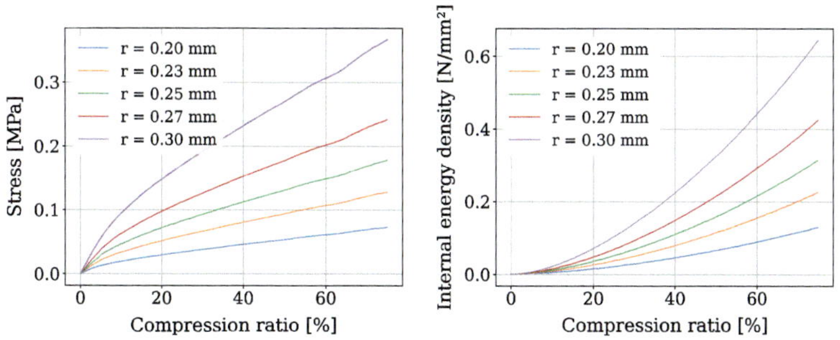

Fig. 5.18 Effect of the material volume density on the structure's stress and internal energy density

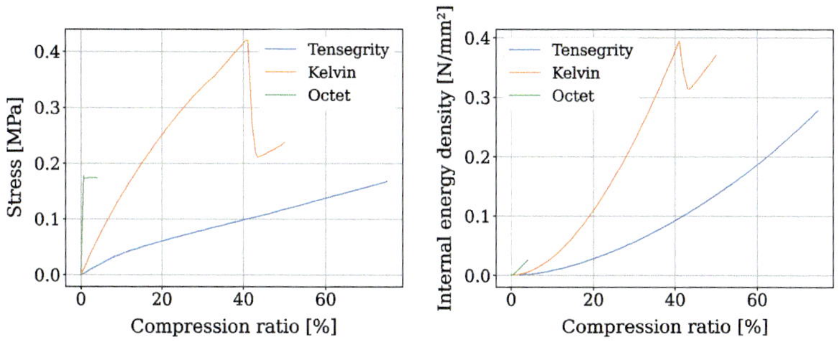

Fig. 5.19 Comparison of the stress-strain and energy density curves of the tensegrity lattice with the octet and Kelvin lattices

Comparison with other lattice structures. For comparison, the same numerical compression tests were performed on two other lattice topologies: the octet and Kelvin lattices ((Bauer et al., 2021)). These tests were performed at the same young's modulus and volume density as the tensegrity lattice. The resulting curves are shown in Fig. 5.19. It can be observed that for the non-tensegrity lattices, localized failure that results in a sudden drop in the stress-strain curve occurs at strains of only a few percents in the octet case and at around 40 % in the Kelvin case. In comparison, the tensegrity lattice remains stable over the whole range of deformation, reaching 75 % strain without failure.

Experimental validation. In addition, experiment test were conducted on structure 3D printed using two-photon polymerization direct laser writing and compared with two other truss-like metamaterials: one based on the octet lattice and the other on the Kelvin lattice (Bauer et al. 2021). Unlike the two other structures, the tensegrity lattice showed an absence of collapse while being compressed, and this up to strains of 80 %. This experimental result is consistent with the numerical simulations presented in this section, and further confirms the stability of tensegrity lattices under uniaxial compression, which can be attributed to the isolated compression members, thus preventing failures of the bars though buckling to propagate.

5.5 Conclusion

In this chapter, we have focused on the study of flexible tensegrity structures, i.e., tensegrity structures operating in a regime in which the assumption of infinitesimal axial deformation of bar members no longer holds. Towards this end, we have presented a reduced order model of compressive members that accounts for buckling in such conditions. This model is a spring mass system whose different parameters are selected to match some of the key features of the equivalent continuous model. These features include the critical buckling force, the inertial properties of the bar, its axial stiffness in the linear regime, and the bending modes of the bar. Regarding buckling, even if only the onset of buckling is considered, the model is able to capture the post-buckling behavior of the bar with higa accuracy in comparison to the buckling load itself. This model enables the efficient simulation of flexible tensegrity systems, either for static or dynamic analysis. It can be used to numerically study the stability of large flexible tensegrity structures and therefore opens the way to the design of new structures with improved mechanical properties.

As a first example, we have used this reduced order model to study the mechanical behavior of a tensegrity lander based on a truncated octahedron shape. It was shown that this lander is capable of storing large amounts of elastic energy during high-velocity impacts, without failure of its members or collapse of the structure. It was also shown that this structure remains stable even after its bars buckle. The buckling of the bars plays an important role in this by capturing the kinetic energy of the lander and limiting the maximum loads in the structure. Avoiding buckling of the bars would result in the structure being restricted to a much lower impact velocity and energy absorption capacity.

Based on these the findings regarding the stability of the truncated octahedron tensegrity for the lander study, we proposed a method to construct tensegrity lattice metamaterials from this tensegrity structure. We have shown through numerical simulations that these tensegrity lattices are stable under large uniaxial compressive strains, unlike other lattice structures such as the octet and Kelvin lattices. Even when all the bars of the structure buckle, the tensegrity lattice remains stable and can withstand strains of up to 75 % without collapsing. This stability seems to be due to the isolated compression paths that are characteristic of the tensegrity lattice. This result was confirmed by experimental tests on 3D printed structures, which showed that the tensegrity lattice remains stable up to strains of 80 %.

By allowing the efficient numerical simulation of flexible tensegrity structures, the reduced order model presented in this work opens the way to understand and design lightweight structures with extreme properties like the presented tensegrity lander and lattice. As for the tensegrity lattice, its capability to withstand large compressive strains without failure makes it a promising candidate for applications in energy absorption and impact absorption. In addition, the separation of the compression and tension members results in unique wave propagation properties that can be tuned by varying the cables prestress (Salahshoor et al., 2018; Pal et al., 2018) and could be used for applications in waveguides or acoustic metamaterials (de Castro Motta et al., 2024).

References

Bauer, J., Kraus, J. A., Crook, C., Rimoli, J. J., & Valdevit, L. (2021). Tensegrity metamaterials: Toward failure-resistant engineering systems through delocalized deformation. *Advanced Materials, 33*(10), 2005647.

Ben Kahla, N., & Kebiche, K. (2000). Nonlinear elastoplastic analysis of tensegrity systems. *Engineering Structures, 22*(11), 1552–1566. ISSN 0141-0296. https://doi.org/10.1016/S0141-0296(99)00088-7, http://www.sciencedirect.com/science/article/pii/S0141029699000887

Courant, R., Friedrichs, K. O., & Lewy, H. (1967). On the partial difference equations of mathematical physics. *IBM Journal of Research and Development, 11*(2), 215–234.

de Castro Motta, J., Garanger, K., & Rimoli, J. J. (2024). Propagation of compression solitary waves on tensegrity-like lattices made of truncated octahedrons. *International Journal of Non-Linear Mechanics, 162*, 104716. https://doi.org/10.1016/j.ijnonlinmec.2024.104716, https://www.sciencedirect.com/science/article/pii/S0020746224000817

Deshpande, V. S., Fleck, N. A., & Ashby, M. F. (2001). Effective properties of the octet-truss lattice material. *Journal of the Mechanics and Physics of Solids, 49*(8), 1747–1769. https://www.sciencedirect.com/science/article/pii/S0022509601000102

Garanger, K., del Valle, I., Rath, M., Krajewski, M., Raheja, U., Pavone, M., & Rimoli, J. J. (2021). Soft tensegrity systems for planetary landing and exploration. In *Earth and Space 2021* (pp. 841–854). https://doi.org/10.1061/9780784483374.078, https://ascelibrary.org/doi/abs/10.1061/9780784483374.078

Gibson, L. J., & Ashby, M. F. (1997). *Cellular Solids: Structure and Properties*. Cambridge Solid State Science Series (2nd ed.). Cambridge University Press. First paperback (1999) edition with corrections.

Johnson, J. B., Bryan, C. W., & Turneaure, F. E. (1905). *The theory and practice of modern framed structures: Designed for the use of schools, and for engineers in professional practice*. Wiley.

Kebiche, K., Kazi-Aoual, M. N., & Motro, R. (1999). Geometrical non-linear analysis of tensegrity systems. *Engineering Structures, 21* (9), 864–876. ISSN 0141-0296. https://doi.org/10.1016/S0141-0296(98)00014-5, http://www.sciencedirect.com/science/article/pii/S0141029698000145

Kraus, J. A., & Rimoli, J. J. (2023). An elastica theory for compressible imperfect beams with application to mechanical metamaterials. *Mechanics Research Communications, 131*, 104147.

Li, Y., Feng, X., Cao, Y., & Gao, H. (2010). Constructing tensegrity structures from one-bar elementary cells. In *Proceedings of the Royal Society of London A: Mathematical, Physical and Engineering Sciences* (Vol. 466, pp. 45–61). The Royal Society. https://royalsocietypublishing.org/doi/10.1098/rspa.2009.0260

Kumar Pal, R., Ruzzene, M., & Rimoli, J. J. (2018). Tunable wave propagation by varying Prestrain in tensegrity-based periodic media. textitExtreme Mechanics Letters, *22*, 149–156. https://doi.org/10.1016/j.eml.2018.06.005, https://www.sciencedirect.com/science/article/pii/S2352431618300993

Rimoli, J. J. (2016). On the impact tolerance of tensegrity-based planetary landers. In *57th AIAA/ASCE/AHS/ASC Structures, Structural Dynamics, and Materials Conference*. https://doi.org/10.2514/6.2016-1511, https://arc.aiaa.org/doi/10.2514/6.2016-1511

Rimoli, J. J. (2018, January). A reduced-order model for the dynamic and post-buckling behavior of tensegrity structures. *Mechanics of Materials, 116*, 146–157. ISSN 0167-6636. https://doi.org/10.1016/j.mechmat.2017.01.009, https://www.sciencedirect.com/science/article/pii/S0167663616304483. IUTAM Symposium on Dynamic Instabilities in Solids.

Rimoli, J. J., & Pal, R. K. (2017, April). Mechanical response of 3-dimensional tensegrity lattices. *Composites Part B: Engineering, 115*, 30–42. ISSN 1359-8368. https://doi.org/10.1016/j.compositesb.2016.10.046, https://www.sciencedirect.com/science/article/pii/S1359836816323538. Composite lattices and multiscale innovative materials and structures.

Salahshoor, H., Kumar Pal, R., & Rimoli, J. J. (2018). Material symmetry phase transitions in three-dimensional tensegrity metamaterials. *Journal of the Mechanics and Physics of Solids, 119*,

382–399. https://doi.org/10.1016/j.jmps.2018.07.011, https://www.sciencedirect.com/science/article/pii/S0022509618304769

SunSpiral, V., Gorospe, G., Bruce, J., Iscen, A., Korbel, G., Milam, S., Agogino, A., & Atkinson, D. (2013). Tensegrity based probes for planetary exploration: Entry, descent and landing (EDL) and surface mobility analysis. In *Proceedings of 10th International Planetary Probe Workshop*. https://www.sunspiral.org/vytas/cv/tensegrity_based_probes.pdf

Zhang, L., Lu, M., Zhang, H. W., & Yan, B. (2015). Geometrically nonlinear elasto-plastic analysis of clustered tensegrity based on the co-rotational approach. *International Journal of Mechanical Sciences, 93*(0), 154–165. ISSN 0020-7403. https://doi.org/10.1016/j.ijmecsci.2015.01.015, http://www.sciencedirect.com/science/article/pii/S0020740315000405

Chapter 6
Tensegrity Lattices with Extremal Mechanical Properties

Anna Al Sabouni-Zawadzka

Abstract This chapter discusses extremal mechanical behavior of tensegrity-inspired metamaterials and structures. It focuses on the identification of extremal properties of modular lattices based on basic tensegrity cells. First, two approaches to the analysis of tensegrity lattices in various scales are described: discrete and continuum, with a special attention paid to the effect of scale when applying the concept of extremal materials to bigger structural systems. Then, two examples of tensegrity lattices are analyzed: a 2D lattice based on hexagonal modules and a 3D lattice based on expanded octahedron. Parameters leading to soft and stiff deformation modes are determined and equivalent mechanical properties, such as Young's and shear moduli and Poisson's ratios, are calculated.

Keywords Tensegrity lattice · Metamaterial · Continuum model · Discrete analysis · Extremal mechanical properties

6.1 Introduction

In recent decades, there has been a significant increase in interest in innovative engineering systems, such as: metamaterials with special properties (Bertoldi et al. 2017; Surjadi et al. 2019; Al Sabouni-Zawadzka et al. 2024), smart materials and structures (Anderson & Sater 2007; Bel Hadj Ali & Smith 2010), adaptive structures (Zawadzki & Al Sabouni-Zawadzka 2020; Lee et al. 2021; Gebara et al. 2022), or extremal materials (Kadic et al., 2012).

The concept of extremal materials was developed by Milton & Cherkaev (1995)—the authors defined them as extremely rigid (stiff) materials in certain modes of deformation and extremely compliant (soft) in others. Apart from this basic feature, which distinguishes extremal materials from traditional ones, they may show various unusual mechanical properties, e.g. a negative Poisson's ratio.

A. Al Sabouni-Zawadzka (✉)
Faculty of Civil Engineering, Warsaw University of Technology, Al. Armii Ludowej 16, 00-637 Warsaw, Poland
e-mail: anna.zawadzka@pw.edu.pl

F. Fraternali and J. J. Rimoli (eds.), *Tensegrity Systems*, CISM International Centre for Mechanical Sciences 2, https://doi.org/10.1007/978-3-031-82283-4_6

Determination of extremal mechanical properties of materials can be performed by analysing the elasticity tensor. After its diagonalization with the use of orthogonal transformation, a set of eigenvalues is determined and orthogonal eigenvectors describe the deformation modes of the considered material. Very small eigenvalues (i.e. approaching zero) indicate a soft material behavior when subjected to deformations described by the corresponding eigenvectors. Based on such an analysis, extremal materials can be classified as follows: nullmode, unimode, bimode, trimode, quadramode, pentamode, and hexamode material (Milton & Cherkaev, 1995; Kadic et al., 2012)—depending on the number of eigenvalues (from zero to six, respectively) of the elasticity tensor which approach zero.

A very important finding described by Milton & Cherkaev (1995) is that extremal materials can be constructed by combining two (or more) specific parent materials—a very compliant and a very rigid one. It suggests that it should be possible to design such a material (i.e. a mechanical metamaterial) and fabricate it, e.g. by using the additive manufacturing—in this case the void space would serve as an extremely soft material and a 3D printed part as an extremely stiff one (Zadpoor, 2016).

Although the concept of extremal materials originally refers to the material scale, it can be also considered in bigger scales. Al Sabouni-Zawadzka (2022) proved that not only materials can exhibit extremal mechanical properties, but also certain structural systems, such as cellular tensegrity-inspired lattices aimed, for example, at civil engineering applications.

This chapter takes a closer look at the analysis of extremal mechanical metamaterials and lattices constructed from basic tensegrity cells. Two approaches (discrete and continuum) to the analysis of such systems are described and a special focus is put on the discussion of scale effects. Then, the identification of extremal mechanical properties of two cellular systems is presented—one in 2D and one in 3D space. Parameters leading to soft and stiff modes of deformations are determined and equivalent mechanical properties, such as Young's modulus or Poisson's ratio, are calculated.

6.2 Employed Notation

This chapter uses the following symbols for key variables: nodal displacement vector ($\mathbf{q}$), nodal force vector ($\mathbf{P}$), linear stiffness matrix ($\mathbf{K}_L$), geometric stiffness matrix ($\mathbf{K}_G$), elasticity matrix ($\mathbf{E}$), cable-to-strut stiffness ratio (k), level of self-stress (σ).

6.3 Discrete Analysis

Tensegrity systems can be analyzed with discrete models, either by using the finite element method (FEM) (Hughes, 2000; Crisfield, 2003) or by direct algebraic formulation (Pelczynski & Gilewski, 2019). Both methods are formally precise and include the influence of prestress.

Tensegrity structures can be described with the relationships:

$$\Delta = \mathbf{B}\mathbf{q}, \quad \omega = \mathbf{C}\mathbf{q}, \tag{6.1}$$

$$\mathbf{N} = \mathbf{D}\Delta, \quad \mathbf{F} = \mathbf{S}\omega, \tag{6.2}$$

$$\mathbf{B}^{\mathrm{T}}\mathbf{N} + \mathbf{C}^{\mathrm{T}}\mathbf{F} = \mathbf{P}, \tag{6.3}$$

where: Δ—elements extension vector (length n), $\mathbf{B}$—compatibility matrix, $\mathbf{q}$—nodal displacement vector (length m), $\boldsymbol{\omega}$—elements rotation vector (length n), $\mathbf{C}$—matrix of rotations, $\mathbf{N}$—vector of normal forces in elements (length n), $\mathbf{F}$—vector of additional generalised forces in elements (length n), $\mathbf{P}$—nodal forces vector (length m),

$$\mathbf{D} = \begin{bmatrix} \frac{E_1 A_1}{L_1} & 0 & \dots & 0 \\ 0 & \frac{E_2 A_2}{L_2} & \dots & 0 \\ \dots & \dots & \dots & \dots \\ 0 & 0 & \dots & \frac{E_n A_n}{L_n} \end{bmatrix}, \quad \mathbf{S} = \begin{bmatrix} \frac{S_1}{L_1} & 0 & \dots & 0 \\ 0 & \frac{S_2}{L_2} & \dots & 0 \\ \dots & \dots & \dots & \dots \\ 0 & 0 & \dots & \frac{S_n}{L_n} \end{bmatrix},$$

E_e, A_e, L_e, S_e—properties of the eth element (Young's modulus, cross-sectional area, length, and prestress force, respectively), $e \in (1, n)$.

Matrices $\mathbf{B}$ and $\mathbf{C}$ can be determined either directly from the structural geometry or by using the FEM formulation, which is more practical in bigger tasks (Al Sabouni-Zawadzka, 2022). Relations (6.1)–(6.3) allow us to obtain a system of displacement-based equations, which considers the nonlinearity of the equations of equilibrium and the influence of self-equilibrated normal forces (further referred to as self-stress) on structural properties:

$$\mathbf{K}\mathbf{q} = \mathbf{P}, \tag{6.4}$$

where: $\mathbf{K} = \mathbf{K}_{\mathrm{L}} + \mathbf{K}_{\mathrm{G}}$—global stiffness matrix, $\mathbf{K}_{\mathrm{L}}$—linear stiffness matrix, $\mathbf{K}_{\mathrm{G}}$—geometric stiffness matrix.

Matrix $\mathbf{K}_{\mathrm{L}}$ can be calculated from the expression $\mathbf{K}_{\mathrm{L}} = \mathbf{B}^{\mathrm{T}}\mathbf{E}\mathbf{B}$ or, following the FEM formalism, by aggregating linear stiffness matrices for single elements $\mathbf{K}_{\mathrm{L}e}$, transformed from the local coordinate system into the global one using generalized Boolean matrices:

$$\mathbf{K}_{\mathrm{L}e} = \frac{E_e A_e}{L_e} \begin{bmatrix} 1 & -1 \\ -1 & 1 \end{bmatrix}, \quad \mathbf{q}_{\mathrm{L}e} = \{u_1, u_2\}, \tag{6.5}$$

$$\mathbf{K}_{\mathrm{L}} = \sum_{e=1}^{n} \mathbf{C}_e^{\mathrm{T}} \mathbf{K}_{\mathrm{L}e} \mathbf{C}_e. \tag{6.6}$$

Matrix $\mathbf{K}_\mathrm{G}$ can be determined from the expression $\mathbf{K}_\mathrm{G} = \mathbf{C}^\mathrm{T}\mathbf{S}\mathbf{C}$ or, similarly to $\mathbf{K}_\mathrm{L}$, by means of the generalised Boolean matrices $\overline{\mathbf{C}}_e$ (dimensions $m \times 6$) and the local element matrix $\mathbf{K}_{\mathrm{G}e}$:

$$\mathbf{K}_{\mathrm{G}e} = \frac{S_e}{L_e}\begin{bmatrix} 0 & 0 & 0 & 0 & 0 & 0 \\ 0 & 1 & 0 & 0 & -1 & 0 \\ 0 & 0 & 1 & 0 & 0 & -1 \\ 0 & 0 & 0 & 0 & 0 & 0 \\ 0 & -1 & 0 & 0 & 1 & 0 \\ 0 & 0 & -1 & 0 & 0 & 1 \end{bmatrix}, \quad \mathbf{q}_{\mathrm{G}e} = \{u_1, v_1, w_1, u_2, v_2, w_2\}, \tag{6.7}$$

$$\mathbf{K}_\mathrm{G} = \sum_{e=1}^{n} \overline{\mathbf{C}}_e^\mathrm{T} \mathbf{K}_{\mathrm{G}e} \overline{\mathbf{C}}_e. \tag{6.8}$$

The number of rows in the Boolean matrices is twice the number of the considered displacements. In $\mathbf{K}_\mathrm{G}$ three displacements (u, v, w) in each node are considered, while in $\mathbf{K}_\mathrm{L}$ only the displacement (u) along the element axis is defined. That is why the matrices $\mathbf{C}_e$ and $\overline{\mathbf{C}}_e$ have different dimensions ($m \times 2$ and $m \times 6$, respectively).

Similarly to traditional trusses, for a statically determinate unsupported tensegrity structure with infinitesimal mechanisms and self-stress states, a generalised Maxwell's truss equation (Calladine, 1978) can be formulated:

$$m - n - r = f - s, \tag{6.9}$$

where: m—number of degrees of freedom, n—number of elements, r—number of required constraints, f—number of infinitesimal mechanisms, s—number of self-stress states.

If we consider a supported structure, the vector of nodal displacements $\tilde{\mathbf{q}}$ has a smaller length of $\tilde{n}$, and the equations of the linear truss mechanics are:

$$\Delta = \tilde{\mathbf{B}}\tilde{\mathbf{q}}, \quad \omega = \tilde{\mathbf{C}}\tilde{\mathbf{q}}, \tag{6.10}$$

$$\mathbf{N} = \mathbf{D}\Delta, \quad \mathbf{F} = \mathbf{S}\omega, \tag{6.11}$$

$$\tilde{\mathbf{B}}^\mathrm{T}\mathbf{N} + \tilde{\mathbf{C}}^\mathrm{T}\mathbf{F} = \tilde{\mathbf{P}}. \tag{6.12}$$

The system of displacement-based equations can be formulated as:

$$\tilde{\mathbf{K}}\tilde{\mathbf{q}} = \tilde{\mathbf{P}}, \tag{6.13}$$

where: $\tilde{\mathbf{K}} = \tilde{\mathbf{K}}_\mathrm{L} + \tilde{\mathbf{K}}_\mathrm{G}$, $\tilde{\mathbf{K}}_\mathrm{L} = \tilde{\mathbf{B}}^\mathrm{T}\mathbf{D}\tilde{\mathbf{B}}$, $\tilde{\mathbf{K}}_\mathrm{G} = \tilde{\mathbf{C}}^\mathrm{T}\mathbf{S}\tilde{\mathbf{C}}$.

The generalised Maxwell's equation takes a form:

$$m - n = f - s. \tag{6.14}$$

6.4 Continuum Analysis

Structural complexity of tensegrity makes them often difficult or even impossible to be analyzed using only discrete methods, especially if tensegrity metamaterials are concerned. In such cases, continuum models allow us to identify equivalent mechanical characteristics of these complex systems (Luo et al., 2008; Kebiche et al., 2008; Rimoli & Pal, 2017).

The continuum model of tensegrity, which is described in this section, was first proposed by Gilewski & Kasprzak (2014) and then developed by Gilewski & Al Sabouni-Zawadzka (2020) and Al Sabouni-Zawadzka (2022). Its application allows us to: identify structural properties corresponding to typical deformation modes (tension, shear), physically interpret the determined mechanical characteristics, estimate the influence of prestress and stiffness of cables and struts on the structural deformation, compare the elastic properties of various tensegrity systems, and identify extremal mechanical properties.

6.4.1 Continuum Model Within the First Gradient Theory

The continuum model is based on the comparison of the strain energy of an unsupported tensegrity truss defined with the use of a discrete model (DM) with the strain energy of a solid according the symmetric linear 3D elasticity theory (LTE) (Green & Zerna, 1968). The strain energy in a discrete model is a quadratic form of nodal displacements $\mathbf{q}$:

$$E_{\mathrm{s}}^{\mathrm{DM}} = \frac{1}{2}\mathbf{q}^{\mathrm{T}}\mathbf{K}\mathbf{q}. \tag{6.15}$$

The strain energy of a solid within the linear theory of elasticity can be expressed as:

$$E_{\mathrm{s}}^{\mathrm{LTE}} = \frac{1}{2}\int\limits_{V} \varepsilon^{\mathrm{T}}\mathbf{E}\varepsilon \, dV, \tag{6.16}$$

where: ε—strain vector, $\mathbf{E}$—elasticity matrix.

Let us consider an unsupported tensegrity structure inscribed into a cube of edge length a and a corresponding solid cube (Fig. 6.1).

Within LTE, the strain energy of a cube of edge length a, assuming that it is constant in the whole volume, equals:

$$E_{\mathrm{s}}^{\mathrm{LTE}} = \frac{1}{2}\varepsilon^{\mathrm{T}}\mathbf{E}\varepsilon a^{3}. \tag{6.17}$$

This assumption is true if the dimensions of the cell are small.

We express nodal displacements by the average mid-values of displacements and their derivatives, using the Taylor series expansion:

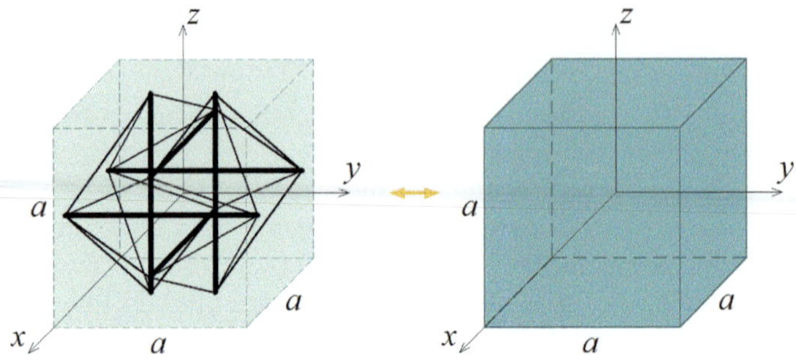

Fig. 6.1 Tensegrity truss and solid cube

$$
\begin{aligned}
& f(x+\Delta x, y+\Delta y, z+\Delta z)=f(x, y, z) \\
& +\frac{\partial f}{\partial x}(x, y, z) \Delta x+\frac{\partial f}{\partial y}(x, y, z) \Delta y+\frac{\partial f}{\partial z}(x, y, z) \Delta z \\
& +\frac{1}{2} \frac{\partial^2 f}{\partial x^2}(x, y, z)(\Delta x)^2+\frac{1}{2} \frac{\partial^2 f}{\partial y^2}(x, y, z)(\Delta y)^2+\frac{1}{2} \frac{\partial^2 f}{\partial z^2}(x, y, z)(\Delta z)^2 \\
& +\frac{\partial^2 f}{\partial x \partial y}(x, y, z) \Delta x \Delta y+\frac{\partial^2 f}{\partial x \partial z}(x, y, z) \Delta x \Delta z+\frac{\partial^2 f}{\partial y \partial z}(x, y, z) \Delta y \Delta z+\ldots .
\end{aligned} \tag{6.18}
$$

Further, the analysis is limited to low order terms (first order approximation), leading to the first gradient theory within LTE. Taken that the coordinates of tensegrity nodes are described by a, for the ith node we obtain: $\{\alpha_{xi}a, \alpha_{yi}a, \alpha_{zi}a\}$. Now, the parameters of the ith node can be expressed by the average mid-values with the following increments: $\Delta x = \alpha_{xi}a$, $\Delta y = \alpha_{yi}a$, $\Delta z = \alpha_{zi}a$.

Comparison of the strain energies (6.15) and (6.17) allows us to determine the equivalent symmetric elasticity matrix:

$$
\mathbf{E} = \begin{bmatrix}
e_{11} & e_{12} & e_{13} & e_{14} & e_{15} & e_{16} \\
 & e_{22} & e_{23} & e_{24} & e_{25} & e_{26} \\
 & & e_{33} & e_{34} & e_{35} & e_{36} \\
 & & & e_{44} & e_{45} & e_{46} \\
 & & & & e_{55} & e_{56} \\
\text{sym.} & & & & & e_{66}
\end{bmatrix}, \tag{6.19}
$$

which, in a general case, has 36 coefficients, including 21 independent ones. Depending on the type of symmetry of the considered task, the elasticity matrix will have a different number of independent coefficients (Chadwick et al., 2001).

Using the continuum model described above, equivalent mechanical properties (technical coefficients) of tensegrity systems can be identified by inverting the elasticity matrix:

$$\mathbf{H} = \mathbf{E}^{-1} = \begin{bmatrix} 1/E_1 & -\nu_{21}/E_2 & -\nu_{31}/E_3 & \lambda_{11}/G_1 & \lambda_{21}/G_2 & \lambda_{31}/G_3 \\ -\nu_{12}/E_1 & 1/E_2 & -\nu_{32}/E_3 & \lambda_{12}/G_1 & \lambda_{22}/G_2 & \lambda_{32}/G_3 \\ -\nu_{13}/E_1 & -\nu_{23}/E_2 & 1/E_3 & \lambda_{13}/G_1 & \lambda_{23}/G_2 & \lambda_{33}/G_3 \\ \kappa_{11}/E_1 & \kappa_{21}/E_2 & \kappa_{31}/E_3 & 1/G_1 & \mu_{21}/G_2 & \mu_{31}/G_3 \\ \kappa_{12}/E_1 & \kappa_{22}/E_2 & \kappa_{32}/E_3 & \mu_{12}/G_1 & 1/G_2 & \mu_{32}/G_3 \\ \kappa_{13}/E_1 & \kappa_{23}/E_2 & \kappa_{33}/E_3 & \mu_{13}/G_1 & \mu_{23}/G_2 & 1/G_3 \end{bmatrix}, \tag{6.20}$$

and using the constitutive LTE relations:

$$\boldsymbol{\sigma} = \mathbf{E}\boldsymbol{\varepsilon}, \quad \boldsymbol{\varepsilon} = \mathbf{E}^{-1}\boldsymbol{\sigma} = \mathbf{H}\boldsymbol{\sigma}. \tag{6.21}$$

where: E_i—Young's moduli, G_i— shear moduli, ν_{ij}—Poisson's ratios, μ_{ij}—relations of shear strains in perpendicular directions, λ_{ij}—relations of normal strains in three directions and shear strain in one direction, κ_{ij}—relations of shear strains in three directions and normal strain in one direction, $i \in (1, 2, 3)$, $j \in (1, 2, 3)$.

Both matrices **E** and **H** have to be positive definite, so the technical coefficients are limited by the following conditions:

$$\begin{aligned} &E_1 > 0, \quad E_2 > 0, \quad E_3 > 0, \quad G_1 > 0, \quad G_2 > 0, \quad G_3 > 0, \\ &\nu_{12}\nu_{21} < 1, \quad \nu_{13}\nu_{31} < 1, \quad \nu_{23}\nu_{32} < 1, \\ &\nu_{12}\nu_{21} + \nu_{13}\nu_{31} + \nu_{23}\nu_{32} + \nu_{12}\nu_{31}\nu_{23} + \nu_{21}\nu_{13}\nu_{32} < 1, \\ &\mu_{12}\mu_{21} < 1, \quad \mu_{13}\mu_{31} < 1, \quad \mu_{23}\mu_{32} < 1, \\ &\mu_{12}\mu_{21} + \mu_{13}\mu_{31} + \mu_{23}\mu_{32} - \mu_{12}\mu_{31}\mu_{23} - \mu_{21}\mu_{13}\mu_{32} < 1. \end{aligned} \tag{6.22}$$

Moreover, principal minors 4×4, 5×5, and 6×6 of matrix **H** have to be positive, what leads to additional global conditions for technical coefficients.

6.4.2 Scale Effects

The continuum model described above is suitable for the analysis of tensegrity systems with the material symmetries, which are acceptable in the linear theory of elasticity. In order to account for higher gradients of displacements, it has to be generalized as shown by Gilewski & Al Sabouni-Zawadzka (2021) and Al Sabouni-Zawadzka (2022). Such a generalised model allows us to assess the scale effect, determine the applicability range of the first approximation theory and identify issues that should be taken into account in bigger scales.

Let us consider a modular lattice constructed from a number of tensegrity-based unit cells (Fig. 6.2). Depending on the unit cell size and its ratio to the dimensions the whole system, including the higher gradients of displacements could lead to different results.

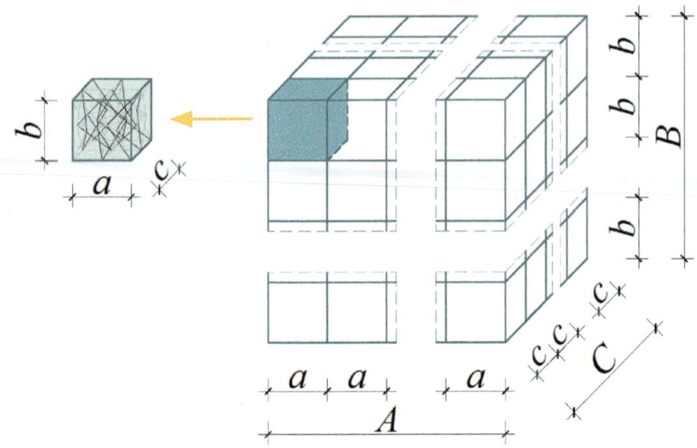

Fig. 6.2 Modular tensegrity lattice

Application of the Taylor series expansion (6.18) for the displacements of all nodes within the unit cell of volume V makes it possible to express the density of the strain energy (6.15) in the form:

$$\tilde{E}_{\mathrm{s}}^{\mathrm{DM}} = \frac{1}{V}[E_{\mathrm{s}11}(\Delta\mathbf{u}) + E_{\mathrm{s}12}(\Delta\mathbf{u}, \Delta\Delta\mathbf{u}) + E_{\mathrm{s}22}(\Delta\Delta\mathbf{u}) + \cdots], \tag{6.23}$$

where: Δ—gradient operator.

The term "11" depends only on the first displacement gradient, the term "22" only on the second displacement gradient, and the term "12" is a mixed term within the theory of elasticity. Let us express those terms as quadratic or bilinear forms:

$$\begin{aligned} E_{\mathrm{s}11}(\Delta\mathbf{u}) &= \frac{1}{2}(\Delta\mathbf{u})^{\mathrm{T}}\mathbf{E}_{11}(\Delta\mathbf{u}), \\ E_{\mathrm{s}12}(\Delta\mathbf{u}, \Delta\Delta\mathbf{u}) &= (\Delta\mathbf{u})^{\mathrm{T}}\mathbf{E}_{12}(\Delta\Delta\mathbf{u}), \\ E_{\mathrm{s}22}(\Delta\Delta\mathbf{u}) &= \frac{1}{2}(\Delta\Delta\mathbf{u})^{\mathrm{T}}\mathbf{E}_{22}(\Delta\Delta\mathbf{u}). \end{aligned} \tag{6.24}$$

Similarly, the energy densities of the higher order gradient theory can be determined. Matrix $\mathbf{E}_{11}$ corresponds to the first approximation within the first gradient theory and is insensitive to scale effects. As it was shown by Al Sabouni-Zawadzka (2022), it can be successfully applied to assess mechanical properties of tensegrity-inspired metamaterials and to identify extremal properties of tensegrity lattices in various scales. Matrices $\mathbf{E}_{12}$ and $\mathbf{E}_{22}$ correspond to the second gradient theory and are sensitive to scale effects. They should be used to analyze mechanical behavior of systems with larger unit cells (relative to the size of the lattice).

6.5 Extremal Mechanical Properties of Tensegrity Lattices

6.5.1 Identification of Extremal Mechanical Properties

Extremal mechanical properties of tensegrity lattices can be identified with the use of discrete (Sect. 6.3) or continuum (Sect. 6.4) approaches. Discrete models are recommended for the analysis of tensegrity systems in a structural scale, while the continuum model is an excellent method for investigating extremal behavior of metamaterials (Al Sabouni-Zawadzka, 2022).

We will use two parameters k and σ:

$$k = \frac{(EA)_{\text{cable}}}{(EA)_{\text{strut}}}, \quad (EA)_{\text{strut}} = EA, \quad \sigma = \frac{S_0}{EA}, \tag{6.25}$$

where: k—cable-to-strut stiffness ratio, σ—level of self-stress, E—Young's modulus of the parent material, A—cross-sectional area of the member, S_0—multiplier of self-stress forces.

To determine extremal mechanical properties of tensegrity lattices in a discrete approach, we perform spectral analysis of the stiffness matrix $\mathbf{K}$:

$$(\mathbf{K}_{\text{L}} + \mathbf{K}_{\text{G}} - \chi\mathbf{I})\,\mathbf{q} = \mathbf{0}. \tag{6.26}$$

Both matrices $\mathbf{K}_{\text{L}}$ and $\mathbf{K}_{\text{G}}$ depend on the parameters k and σ. Zero eigenvalues χ (excluding the values corresponding to rigid movements) indicate soft modes of deformation, while positive values of χ correspond to medium and stiff deformation modes. Soft modes are described by the eigenvectors $\mathbf{q}_{\text{soft}}$, which generate zero value of the strain energy:

$$\frac{1}{2}\mathbf{q}_{\text{soft}}^{\text{T}}\,(\mathbf{K}_{\text{L}} + \mathbf{K}_{\text{G}})\,\mathbf{q}_{\text{soft}} = 0. \tag{6.27}$$

At the end, we find pairs of parameters k_{extr} and σ_{extr}, for which soft, medium, and stiff modes are obtained. Soft modes indicate an extremely compliant behavior of the structure in certain deformation states, while stiff modes correspond to extremely large stiffness of the system in other states.

The main disadvantage of the discrete approach is that it leads to very big dimensions of stiffness matrices, and even in simple tasks the spectral analysis involves high computational costs. This problem does not occur in the continuum approach, which is suitable for the determination of extremal mechanical properties of tensegrity-based metamaterials.

The continuum approach is based on the spectral analysis of the elasticity matrix $\mathbf{E}$ determined using the continuum model described in Sect. 6.4. Matrix $\mathbf{E}$ has dimensions of 3×3 (in 2D space) or 6×6 (in 3D space), independent of the task complexity, and it includes two parameters: k and σ. After solving the eigenvalue problem:

$$(\mathbf{E} - \lambda\mathbf{I})\,\mathbf{w} = \mathbf{0}, \tag{6.28}$$

it is possible to determine the values of parameters k_{extr} and σ_{extr}, which ensure occurrence of soft, medium, and stiff modes of deformation—zero eigenvalues λ (excluding the values corresponding to rigid movements) indicate soft deformation modes, medium and stiff modes correspond to the positive values of λ. Deformation states are described by the corresponding eigenvectors $\mathbf{w}$—starting from soft modes in which the system is extremely compliant, up to stiff modes where the structure is extremely stiff.

The procedure based on the continuum approach is illustrated below using two examples of tensegrity lattices—one in 2D and one in 3D space. Both lattices are modular systems based on basic tensegrity modules—unit cells—connected in nodes to form a supercell—a repetitive unit that can be used to create a metamaterial of any volume. The hexagon-based supercell (2D example) is a regular system, in which all modules are oriented in the same direction. As was proved by Al Sabouni-Zawadzka (2022), in such systems the properties of the supercell are the same as the properties of the module. The expanded octahedron-based supercell (3D example) is created from modules rotated clockwise and counter-clockwise connected in such a way that they form a system with three symmetry planes. In this case, the supercell has different properties than the single module (Al Sabouni-Zawadzka, 2022).

6.5.2 Example 1—Tensegrity Lattice in 2D Space

The first example is a 2D tensegrity lattice consisting of four hexagonal modules (unit cells) connected in nodes. Geometry of the module and the modular lattice is presented in Fig. 6.3.

Hexagonal module consists of six cables and three struts, and has one infinitesimal mechanism and one corresponding self-stress state: $S_1 = S_3 = S_4 = S_6 = 2/\sqrt{3} \cdot S_0$, $S_2 = S_5 = 1/\sqrt{3} \cdot S_0$, $S_7 = -2/\sqrt{3} \cdot S_0$, $S_8 = S_9 = S_0$ (element numbering according to Fig. 6.3).

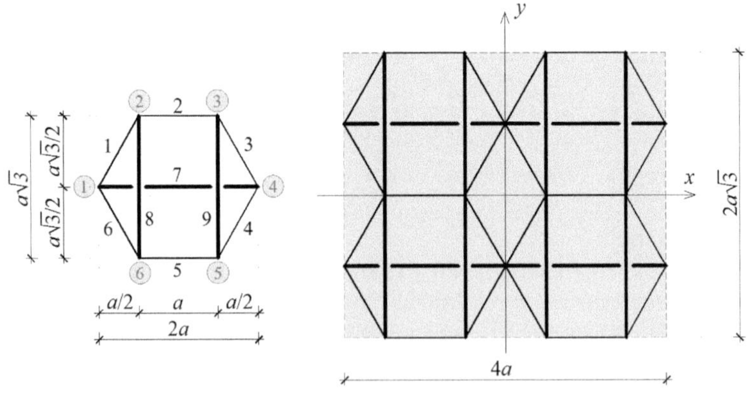

Fig. 6.3 Hexagonal module (unit cell) and modular lattice (supercell) inscribed into a rectangle

We build the continuum model for the hexagon-based supercell and obtain the elasticity matrix in a general form:

$$\mathbf{E}_{\text{hexa}} = \begin{bmatrix} e_{11} & e_{12} & 0 \\ e_{12} & e_{22} & 0 \\ 0 & 0 & e_{33} \end{bmatrix}, \tag{6.29}$$

where:

$$e_{11} = \frac{9k + 2\sqrt{3}\sigma + 8}{8\sqrt{3}}, \quad e_{12} = \frac{\sqrt{3}k - 2\sigma}{8}, \quad e_{22} = \frac{3\sqrt{3}k + 2\sigma + 8}{8},$$

$$e_{33} = \frac{\sqrt{3}k - 2\sigma}{8}.$$

After substituting its coefficients, matrix $\mathbf{E}_{\text{hexa}}$ can be presented as:

$$\mathbf{E}_{\text{hexa}} = \frac{EA}{ah}\frac{1}{8\sqrt{3}}\begin{bmatrix} 8+9k & 3k & 0 \\ 3k & 8\sqrt{3}+9k & 0 \\ 0 & 0 & 3k \end{bmatrix} + \frac{S_0}{ah}\frac{1}{4}\begin{bmatrix} 1 & -1 & 0 \\ -1 & 1 & 0 \\ 0 & 0 & -1 \end{bmatrix}. \tag{6.30}$$

From the spectral analysis of the matrix $\mathbf{E}_{\text{hexa}}$ we determine eigenvalues, which describe extremal properties of the module:

$$\begin{aligned}
\lambda_1 &= \frac{EA}{24ah}\left(\begin{array}{l} 12 + 4\sqrt{3} + 9\sqrt{3}k + 6\sigma \\ + \sqrt{96\left(2-\sqrt{3}\right) + 27k^2 - 36\sqrt{3}k\sigma + 36\sigma^2} \end{array}\right) > 0, \\
\lambda_2 &= \frac{EA}{24ah}\left(\begin{array}{l} 12 + 4\sqrt{3} + 9\sqrt{3}k + 6\sigma \\ - \sqrt{96\left(2-\sqrt{3}\right) + 27k^2 - 36\sqrt{3}k\sigma + 36\sigma^2} \end{array}\right) > 0, \\
\lambda_3 &= \frac{EA}{8ah}\left(\sqrt{3}k - 2\sigma\right).
\end{aligned} \tag{6.31}$$

By comparing the last eigenvalue to zero, we obtain the relation between the parameters k_{extr} and σ_{extr}, which ensures occurrence of the soft mode of deformation:

$$\sigma_{\text{extr}} = \frac{k_{\text{extr}}\sqrt{3}}{2}. \tag{6.32}$$

After substituting relation (6.32) into (6.31), we get the following eigenvalues:

$$\lambda_1 = \frac{EA}{2ah}(\sqrt{3}\cdot k_{\text{extr}} + 2), \quad \lambda_2 = \frac{EA}{2ah\sqrt{3}}(3\cdot k_{\text{extr}} + 2), \quad \lambda_3 = 0. \tag{6.33}$$

Eigenvectors corresponding to these eigenvalues describe deformation states of the lattice. Two stiff modes of deformation can be identified: the first one (Fig. 6.4a),

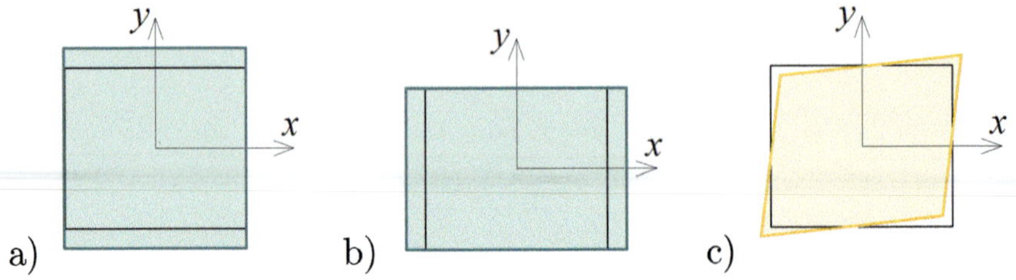

$$\mathbf{w}_{1,\mathrm{hexa}} = \begin{bmatrix} 0 \\ 1 \\ 0 \end{bmatrix} \qquad \mathbf{w}_{2,\mathrm{hexa}} = \begin{bmatrix} 1 \\ 0 \\ 0 \end{bmatrix} \qquad \mathbf{w}_{3,\mathrm{hexa}} = \begin{bmatrix} 0 \\ 0 \\ 1 \end{bmatrix}$$

Fig. 6.4 Stiff and soft deformation modes of hexagonal lattice: **a** stiff mode (λ_1); **b** stiff mode (λ_2); **c** soft mode (λ_3)

represented by the eigenvector $\mathbf{w}_{1,\mathrm{hexa}}$, is volumetric with an extension along axis y, the second one (Fig. 6.4b), represented by the eigenvector $\mathbf{w}_{2,\mathrm{hexa}}$, is also volumetric with an extension along axis x. Additionally, one soft mode of deformation (Fig. 6.4c) is recognized. It is represented by the eigenvector $\mathbf{w}_{3,\mathrm{hexa}}$ and it is a shear deformation in plane x–y .

The next step of the analysis is the calculation of equivalent mechanical properties (technical coefficients) of the lattice. In order to do it, we invert the elasticity matrix $\mathbf{E}_{\mathrm{hexa}}$ (6.29) and compare it to matrix $\mathbf{H}$ (6.20) (reduced to 2D task), which contains the technical coefficients that we want to determine:

$$\mathbf{E}_{\mathrm{hexa}}^{-1} = \begin{bmatrix} \frac{e_{22}}{e_{11}e_{22} - e_{12}^2} & \frac{e_{12}}{e_{12}^2 - e_{11}e_{22}} & 0 \\ \frac{e_{12}}{e_{12}^2 - e_{11}e_{22}} & \frac{e_{11}}{e_{11}e_{22} - e_{12}^2} & 0 \\ 0 & 0 & \frac{1}{e_{33}} \end{bmatrix} = \begin{bmatrix} \frac{1}{E_1} & -\frac{\nu_{21}}{E_2} & 0 \\ -\frac{\nu_{12}}{E_1} & \frac{1}{E_2} & 0 \\ 0 & 0 & \frac{1}{G_1} \end{bmatrix} = \tag{6.34}$$
$$= \mathbf{H}_{\mathrm{hexa}}.$$

Now we can determine the formulas for technical coefficients:

$$E_1 = e_{11} - \frac{e_{12}^2}{e_{22}}, \quad E_2 = e_{22} - \frac{e_{12}^2}{e_{11}}, \quad G_1 = e_{33}, \quad \nu_{12} = \frac{e_{12}}{e_{22}}, \quad \nu_{21} = \frac{e_{12}}{e_{11}}. \tag{6.35}$$

The conditions limiting the values of technical coefficients can be obtained from the symmetry of the matrix $\mathbf{H}_{\mathrm{hexa}}$ and the fact that the matrices $\mathbf{E}_{\mathrm{hexa}}$ and $\mathbf{H}_{\mathrm{hexa}}$ need to be positive definite:

$$E_1 > 0, \quad E_2 > 0, \quad G_1 > 0, \quad \nu_{12}\nu_{21} < 1, \quad \frac{\nu_{12}}{E_1} = \frac{\nu_{21}}{E_2}. \tag{6.36}$$

From the limiting conditions we can determine the domain for all technical coefficients:

$$\sigma < \frac{k\sqrt{3}}{2}. \tag{6.37}$$

Figures 6.5, 6.6 and 6.7 show plots of technical coefficients as a function of two parameters k and σ. The graphs clearly show that the values of all determined

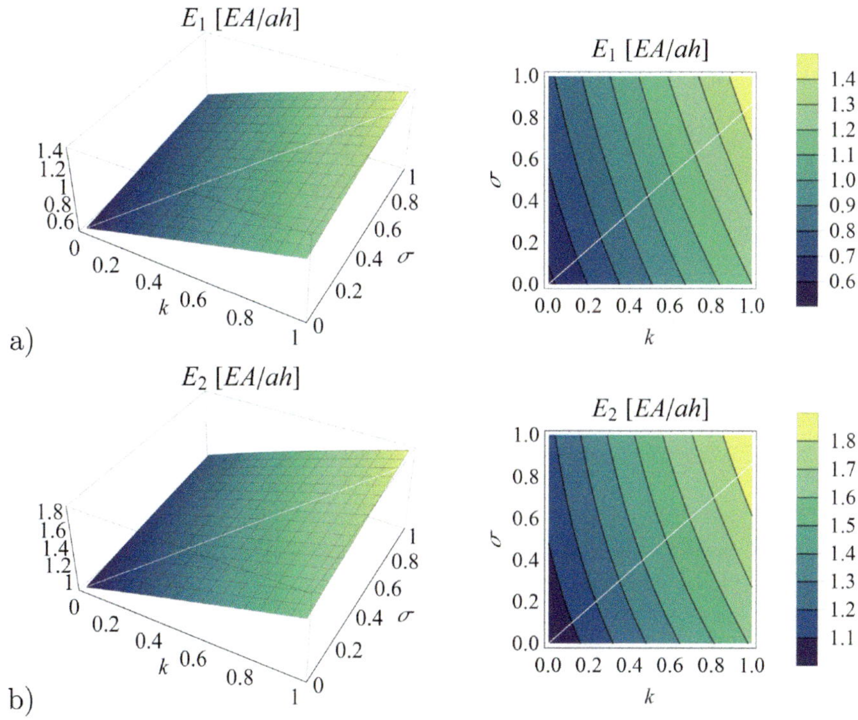

Fig. 6.5 Hexagonal lattice: **a** Young's modulus $E_1 > 0$; **b** Young's modulus $E_2 > 0$

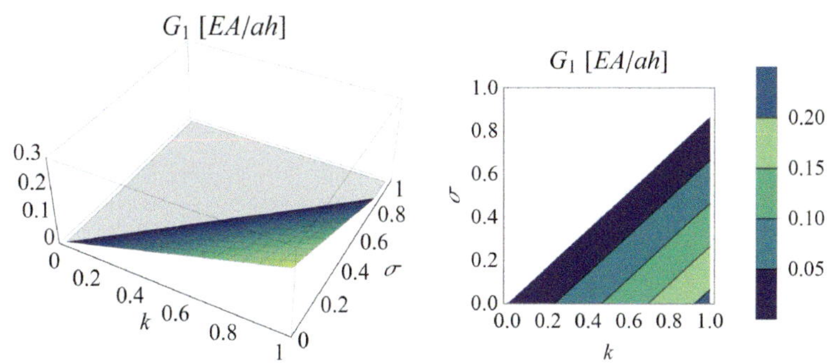

Fig. 6.6 Hexagonal lattice—shear modulus $G_1 > 0$

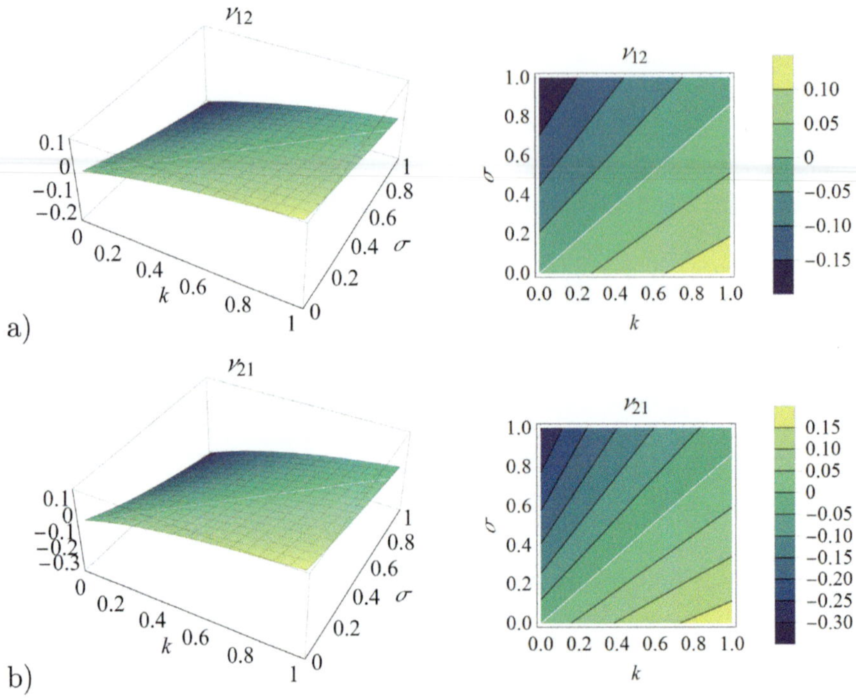

Fig. 6.7 Hexagonal lattice: **a** Poisson's ratio ν_{12}; **b** Poisson's ratio ν_{21}

coefficients can be controlled by changing the stiffness of cables and struts, or by adjusting the level of self-stress.

Let us check how the lattice behaves in the extremal state. It can be noticed that the domain of the technical coefficients (6.37) is bounded by the function that is the same as the relation (6.32) that leads to extremal mechanical properties of the hexagonal lattice. Therefore, it is not possible to find the values of technical coefficients that would correspond to the determined soft mode of deformation. However, we can estimate these values for the pairs of parameters $\{k, \sigma\} \to \{k_{\text{extr}}, \sigma_{\text{extr}}\}$.

Let us assume $\{k_{\text{extr}}, \sigma_{\text{extr}}\} = \{0.1, 0.0866025\}$—the pair fulfills the relation (6.32). We obtain:

$$
\begin{aligned}
&E_1 \to 0.6640 \cdot EA/ah, \\
&E_2 \to 1.0866 \cdot EA/ah, \\
&G_1 \to 0, \\
&\nu_{12} \to 0, \quad \nu_{21} \to 0.
\end{aligned}
\tag{6.38}
$$

The technical coefficients show that the lattice is stiff under the compression in both directions and compliant under shear, which is consistent with the identified deformation modes (Fig. 6.4).

6.5.3 Example 2—Tensegrity Lattice in 3D Space

The second example is a 3D tensegrity lattice consisting of eight expanded octahedron modules (unit cells) connected in nodes. Geometry of the module and the modular lattice is presented in Fig. 6.8 and in Tab. 6.1.

Expanded octahedron module consists of twenty four cables and six struts, and has one infinitesimal mechanism and one corresponding self-stress state: $S_{\text{strut}} = -S_0$, $S_{\text{cable}} = 0.408247 \cdot S_0$—all cables (and all struts) have the same length and are prestressed with the same value of the force.

We build the continuum model for the expanded octahedron-based supercell and obtain the elasticity matrix:

$$\mathbf{E}_{\text{octa}} = \begin{bmatrix} e_{11} & e_{12} & e_{13} & 0 & 0 & 0 \\ & e_{11} & e_{13} & 0 & 0 & 0 \\ & & e_{33} & 0 & 0 & 0 \\ & & & e_{12} & 0 & 0 \\ & & & & e_{13} & 0 \\ \text{sym.} & & & & & e_{13} \end{bmatrix}, \tag{6.39}$$

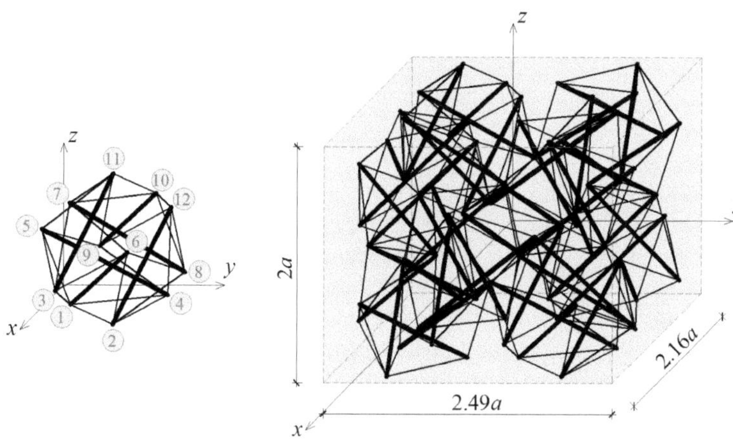

Fig. 6.8 Expanded octahedron module (unit cell) and modular lattice (supercell) inscribed into a cuboid

Table 6.1 Nodal coordinates of the expanded octahedron module

Node	x	y	z	Node	x	y	z
1	0.849*a*	0.356*a*	0	7	0.926*a*	0.757*a*	*a*
2	0.617*a*	1.025*a*	0	8	0	0.935*a*	0.333*a*
3	1.080*a*	0.935*a*	0.333*a*	9	0.540*a*	0	0.333*a*
4	0.154*a*	0.490*a*	0	10	0.231*a*	0.891*a*	*a*
5	1.080*a*	0.312*a*	0.667*a*	11	0.463*a*	0.223*a*	*a*
6	0.540*a*	1.247*a*	0.667*a*	12	0	0.311*a*	0.667*a*

where: $e_{11} = 2EA/a^2$ $(1.31223 \cdot k - 0.033882 \cdot \sigma + 0.428571)$, $e_{12} = EA/a^2$ $(0.874818 \cdot -0.0225873 \cdot \sigma + 0.285714)$, $e_{13} = EA/a^2$ $(0.699854 \cdot k + 0.0903512 \cdot \sigma + 0.571429)$, $e_{33} = 2EA/a^2$ $(1.39971 \cdot k - 0.0903512 \cdot \sigma + 0.285714)$.

From the spectral analysis of matrix $\mathbf{E}_{\text{octa}}$ we can find pairs of parameters k_{extr} and σ_{extr}, which lead to extremal behavior of the lattice. Let us assume $k_{\text{extr}} = 0.1$. Then, in order to obtain a zero eigenvalue of the elasticity matrix, we need to adopt $\sigma_{\text{extr}} = 0.774593$. With the established values of k_{extr} and σ_{extr}, we obtain the following eigenvalues and corresponding eigenvectors, which describe extremal mechanical properties of the lattice:

$$
\begin{aligned}
\lambda_1 &= 2.134200\frac{EA}{a^2}: & \mathbf{w}_{1,\text{octa}} &= [1 \quad 1 \quad 1 \quad 0 \quad 0 \quad 0]^{\text{T}},\\
\lambda_2 &= 0.711399\frac{EA}{a^2}: & \mathbf{w}_{2,\text{octa}} &= [0 \quad 0 \quad 0 \quad 0 \quad 1 \quad 0]^{\text{T}},\\
\lambda_3 &= 0.711399\frac{EA}{a^2}: & \mathbf{w}_{3,\text{octa}} &= [0 \quad 0 \quad 0 \quad 0 \quad 0 \quad 1]^{\text{T}},\\
\lambda_4 &= 0.711399\frac{EA}{a^2}: & \mathbf{w}_{4,\text{octa}} &= [1 \quad -1 \quad 0 \quad 0 \quad 0 \quad 0]^{\text{T}},\\
\lambda_5 &= 0.355700\frac{EA}{a^2}: & \mathbf{w}_{5,\text{octa}} &= [0 \quad 0 \quad 0 \quad 1 \quad 0 \quad 0]^{\text{T}},\\
\lambda_6 &= 0: & \mathbf{w}_{6,\text{octa}} &= [0.5 \quad 0.5 \quad -1 \quad 0 \quad 0 \quad 0]^{\text{T}}.
\end{aligned}
\tag{6.40}
$$

Deformation states of the lattice, described by the determined eigenvectors, are presented in Fig. 6.9. One stiff mode of deformation can be identified (Fig. 6.9a), which is volumetric with equal extensions along three axes. Moreover, there are four medium modes: shear deformation in plane x–z (Fig. 6.9b), shear deformation in plane y–z (Fig. 6.9c), volumetric deformation with an extension along axis x and a contraction along axis y (Fig. 6.9d), and shear deformation in plane x–y (Fig. 6.9e). In the analyzed lattice, there is one soft mode of deformation (Fig. 6.9f), which is volumetric with a contraction along axis z and extensions along axes x and y.

Let us now calculate equivalent mechanical properties (technical coefficients) of the lattice. We invert the elasticity matrix $\mathbf{E}_{\text{octa}}$ (6.39):

$$
\mathbf{E}_{\text{octa}}^{-1} = \mathbf{H}_{\text{octa}} = \begin{bmatrix}
\frac{1}{E_1} & -\frac{\nu_{12}}{E_1} & -\frac{\nu_{31}}{E_3} & 0 & 0 & 0\\
-\frac{\nu_{12}}{E_1} & \frac{1}{E_1} & -\frac{\nu_{31}}{E_3} & 0 & 0 & 0\\
-\frac{\nu_{13}}{E_1} & -\frac{\nu_{13}}{E_1} & \frac{1}{E_3} & 0 & 0 & 0\\
0 & 0 & 0 & \frac{1}{G_1} & 0 & 0\\
0 & 0 & 0 & 0 & \frac{1}{G_2} & 0\\
0 & 0 & 0 & 0 & 0 & \frac{1}{G_2}
\end{bmatrix},
\tag{6.41}
$$

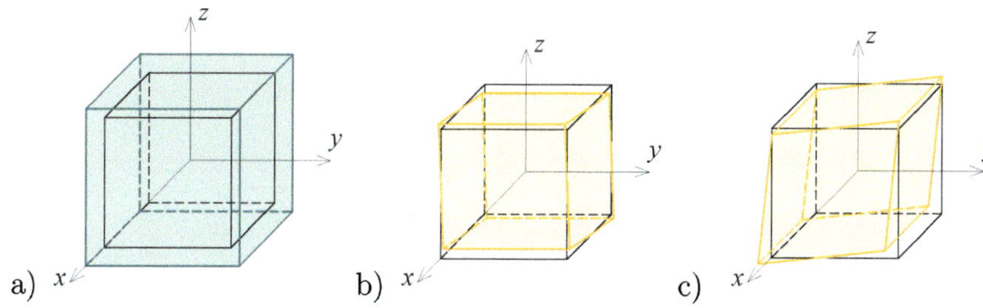

$$\mathbf{w}_{1,\mathrm{octa}} = \begin{bmatrix} 1 \\ 1 \\ 1 \\ 0 \\ 0 \\ 0 \end{bmatrix} \qquad \mathbf{w}_{2,\mathrm{octa}} = \begin{bmatrix} 0 \\ 0 \\ 0 \\ 0 \\ 1 \\ 0 \end{bmatrix} \qquad \mathbf{w}_{3,\mathrm{octa}} = \begin{bmatrix} 0 \\ 0 \\ 0 \\ 0 \\ 0 \\ 1 \end{bmatrix}$$

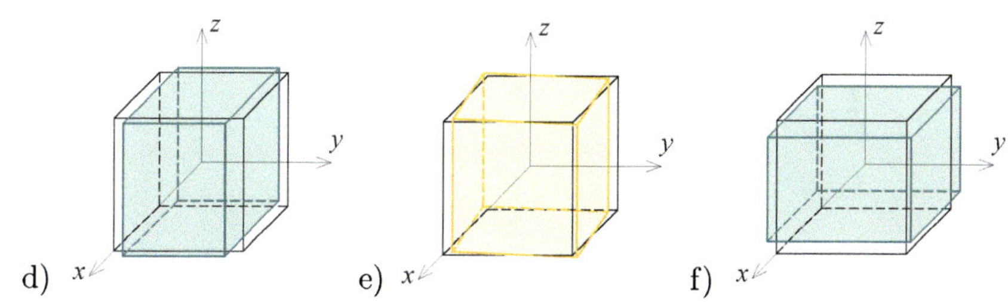

$$\mathbf{w}_{4,\mathrm{octa}} = \begin{bmatrix} 1 \\ -1 \\ 0 \\ 0 \\ 0 \\ 0 \end{bmatrix} \qquad \mathbf{w}_{5,\mathrm{octa}} = \begin{bmatrix} 0 \\ 0 \\ 0 \\ 1 \\ 0 \\ 0 \end{bmatrix} \qquad \mathbf{w}_{6,\mathrm{octa}} = \begin{bmatrix} 0.5 \\ 0.5 \\ -1 \\ 0 \\ 0 \\ 0 \end{bmatrix}$$

Fig. 6.9 Deformation modes of expanded octahedron lattice: **a** stiff mode corresp. to λ_1; **b** medium mode corresp. to λ_2; **c** medium mode corresp. to λ_3; **d** medium mode corresp. to λ_4; **e** medium mode corresp. to λ_5; **f** soft mode corresp. to λ_6

and determine the formulas for technical coefficients:

$$E_1 = E_2 = \frac{(e_{11} - e_{12})\left(e_{33}(e_{11} + e_{12}) - 2e_{13}^2\right)}{e_{11}e_{33} - e_{13}^2}, \quad E_3 = \frac{e_{33}(e_{11} + e_{12}) - 2e_{13}^2}{e_{11} + e_{12}},$$
$$G_1 = e_{12}, \quad G_2 = G_3 = e_{13},$$
$$\nu_{12} = \nu_{21} = \frac{e_{13}^2 - e_{12}e_{33}}{e_{13}^2 - e_{11}e_{33}}, \quad \nu_{13} = \nu_{23} = \frac{e_{13}(e_{11} - e_{12})}{e_{11}e_{33} - e_{13}^2}, \quad \nu_{31} = \nu_{32} = \frac{e_{13}}{e_{11} + e_{12}}. \tag{6.42}$$

The conditions limiting the values of technical coefficients can be obtained from the symmetry of the matrix $\mathbf{H}_{\text{octa}}$ and the fact that the matrices $\mathbf{E}_{\text{octa}}$ and $\mathbf{H}_{\text{octa}}$ need to be positive definite:

$$\begin{gathered} E_1 > 0, \quad E_2 > 0, \quad E_3 > 0, \quad G_1 > 0, \quad G_2 > 0, \quad G_3 > 0, \\ \nu_{12}\nu_{21} < 1, \quad \nu_{13}\nu_{31} < 1, \quad \nu_{23}\nu_{32} < 1, \quad \nu_{13}/E_1 = \nu_{31}/E_3, \\ \nu_{12}\nu_{21} + \nu_{13}\nu_{31} + \nu_{23}\nu_{32} + \nu_{12}\nu_{31}\nu_{23} + \nu_{21}\nu_{13}\nu_{32} < 1. \end{gathered} \tag{6.43}$$

From the limiting conditions:

$$\sigma < 7.74593 \cdot k. \tag{6.44}$$

Figures 6.10, 6.11, 6.12, 6.13, 6.14, 6.15 and 6.16 show plots of technical coefficients as a function of two parameters k and σ. Similarly to the previously analyzed hexagonal lattice, it can be noticed that the values of all determined coefficients

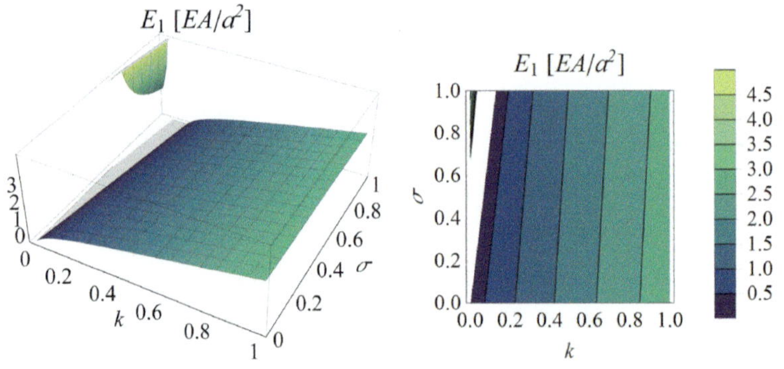

Fig. 6.10 Expanded octahedron lattice—Young's modulus $E_1 > 0$

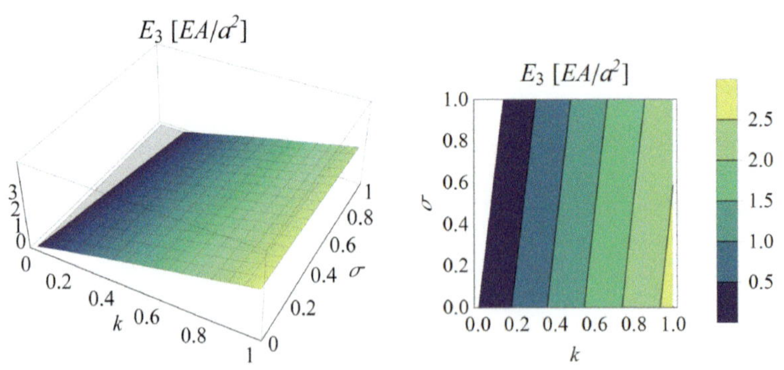

Fig. 6.11 Expanded octahedron lattice—Young's modulus $E_3 > 0$

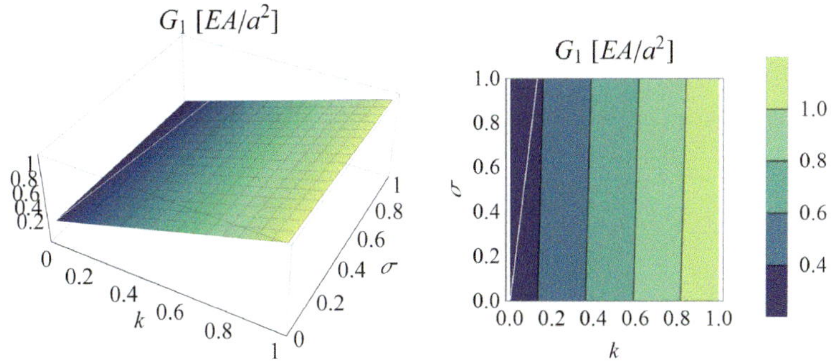

Fig. 6.12 Expanded octahedron lattice—shear modulus $G_1 > 0$

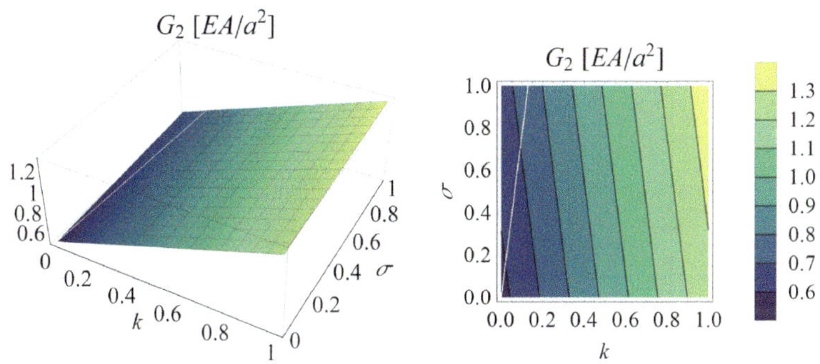

Fig. 6.13 Expanded octahedron lattice—shear modulus $G_2 > 0$

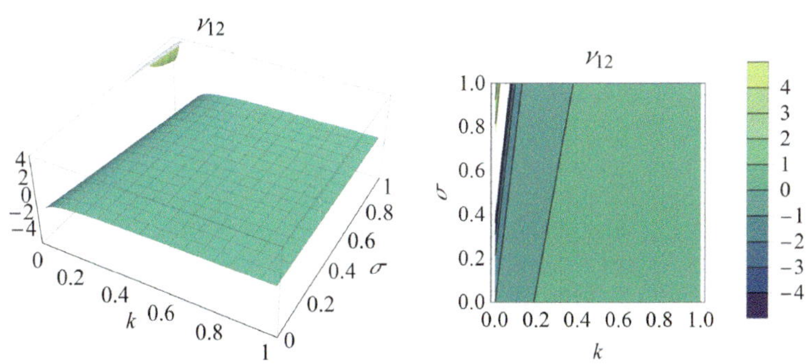

Fig. 6.14 Expanded octahedron lattice—Poisson's ratio ν_{12}

can be controlled by changing the stiffness of cables and struts, or by adjusting the self-stress level.

In order to check how the lattice behaves in the extremal state, let us estimate the values of technical coefficients for $\{k, \sigma\} \to \{k_{\text{extr}}, \sigma_{\text{extr}}\} = \{0.1, 0.774593\}$:

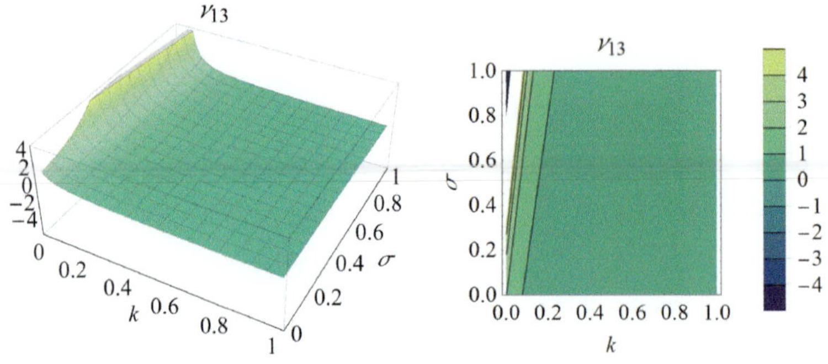

Fig. 6.15 Expanded octahedron lattice—Poisson's ratio ν_{13}

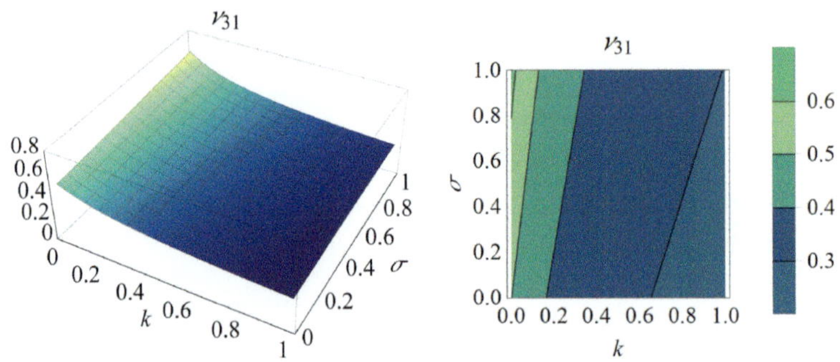

Fig. 6.16 Expanded octahedron lattice—Poisson's ratio ν_{31}

$$
\begin{aligned}
&E_1 = E_2 \to 0, \quad E_3 \to 0, \\
&G_1 \to 0.3557 \cdot EA/a^2, \quad G_2 = G_3 \to 0.7114 \cdot EA/a^2, \\
&\nu_{12} = \nu_{21} \to -1, \quad \nu_{13} = \nu_{23} \to 2, \quad \nu_{31} = \nu_{32} \to 0.5.
\end{aligned}
\tag{6.45}
$$

The technical coefficients show that the lattice is compliant in the volumetric deformation states (Young's moduli approaching zero) and stiff under shear (positive shear moduli). Moreover, one of the Poisson's ratios reaches a negative value, which is one of the desired characteristics of extremal metamaterials.

6.6 Concluding Remarks

In this chapter, the concept of extremal materials was applied to cellular tensegrity lattices. Two systems were analyzed: a 2D lattice based on hexagonal modules and a 3D lattice based on expanded octahedron modules. For each metamaterial, a thorough identification of extremal mechanical properties was performed. First, a continuum model was built, in order to determine the elasticity matrix **E**. Then, a spectral analysis of the matrix was performed, and eigenvalues and eigenvectors describing soft and stiff deformations modes were found. The next step was the determination of equivalent mechanical properties of the lattices, based on the continuum model. Graphs of technical coefficients, such as Young's moduli, shear moduli, and Poisson's ratios, showed that the values of all determined coefficients can be adjusted by controlling the cable-to-strut stiffness ratio and the self-stress level. The analysis revealed that both structures have only one soft mode of deformation—in the hexagonal lattice it is a shear deformation and in the expanded octahedron system—a volumetric one.

Development of systems with extremal modes of deformation.

Further research will focus on the identification of extremal mechanical properties of other configurations of tensegrity-based lattices. Systems constructed from other tensegrity modules and with different patterns of unit cells will be considered. Moreover, metamaterials with graded mechanical properties will be analyzed.

References

Al Sabouni-Zawadzka, A. (2022). *High performance tensegrity-inspired metamaterials and structures*. CRC Press.

Al Sabouni-Zawadzka, A., Gilewski, W., Nazifi Charandabi, R., & Zawadzki, A. (2024). Stability of tensegrity-inspired structures fabricated through additive manufacturing. *Composite Structures, 345*.

Anderson, E. H., & Sater, J. M. (2007). SPIE smart structures product implementation award: A review of the first ten Years. In *Industrial and Commercial Applications of Smart Structures Technologies 2007* (Vol. 6527, pp. 1—13).

Bel Hadj Ali, N., & Smith, I. (2010) Dynamic behavior and vibration control of a tensegrity structure. *International Journal of Solids and Structures, 47*(9).

Bertoldi, K., Vitelli, V., Christensen, J., & Hecke, M. (2017). Flexible mechanical metamaterials. *Nature Reviews Materials, 2*(11).

Calladine, C. R. (1978). Buckminster Fuller's "Tensegrity" structures and Clerk Maxwell's rules for the construction of stiff frames. *International Journal of Solids and Structures, 14*(2), 161–172.

Chadwick, P., Vianello, M., & Cowin, S. (2001). A new proof that the number of linear elastic symmetries is eight. *Journal of the Mechanics and Physics of Solids, 49*(11), 2471–2492.

Crisfield, M. A. (2003). *Non-linear finite element analysis of solids and structures*. Volume 1: Essentials. Wiley.

Gebara, C., Lytal, P., & Rimoli, J. (2022). Additive manufacturing of compliant mechanisms for deployable aerospace structures. *Journal of Materials Engineering and Performance, 31*.

Gilewski, W., & Al Sabouni-Zawadzka, A. (2020). Equivalent mechanical properties of tensegrity truss structures with self-stress included. *European Journal of Mechanics—A/Solids, 83*(4).

Gilewski, W., & Al Sabouni-Zawadzka, A. (2021). Towards recognition of scale effects in a solid model of lattices with tensegrity-inspired microstructure. *Solids, 2*(1), 50–59.

Gilewski, W., & Kasprzak, A. (2014). 3D continuum models of tensegrity modules with the effect of self-stress. In *11th World Congress on Computational Mechanics, 5th European Conference on Computational Mechanics, Barcelona*.

Green, A. E., & Zerna, W. (1968). *Theoretical elasticity*. Oxford, UK: Oxford University Press.

Hughes, T. (2000). *the finite element method: linear static and dynamic finite element analysis* (vol. 78). Dover Publications.

Kadic, M., Bückmann, T., Stenger, N., Thiel, M., & Wegener, M. (2012). On the practicability of pentamode mechanical metamaterials. *Applied Physics Letters, 100*(19), 191901.

Kebiche, K., Aoual, K., & Motro, R. (2008). Continuum models for systems in a selfstress state. *International Journal of Space Structures, 23*(2), 103–115.

Lee, T. U., Chen, Y., Heitzmann, M. T., & Gattas, J. M. (2021). Compliant curved-crease origami-inspired metamaterials with a programmable force-displacement response. *Materials & Design, 207*(40).

Luo, Y., Xu, X., Lele, T., Kumar, S., & Ingber, D. E. (2008). A multi-modular tensegrity model of an actin stress fiber. *Journal of Biomechanics, 41*(11), 2379–2387.

Milton, G., & Cherkaev, A. (1995). Which elasticity tensors are realizable? *Journal of Engineering Materials and Technology, 117*.

Pelczynski, J., & Gilewski, W. (2019). Algebraic formulation for moderately thick elastic frames, beams, trusses, and grillages within Timoshenko theory. *Mathematical Problems in Engineering, 2019*(1), 9.

Rimoli, J., & Pal, R. K. (2017). Mechanical response of 3-dimensional tensegrity lattices. *Composites Part B: Engineering, 115*, 30–42.

Surjadi, J. U., Gao, L., Du, H., Li, X., Xiong, X., Fang, N. X., & Lu, Y. (2019). Mechanical metamaterials and their engineering applications. *Advanced Engineering Materials, 21*(3).

Zadpoor, A. A. (2016). Mechanical meta-materials. *Materials Horizons, 3*(5), 371–381.

Zawadzki, A., & Al Sabouni-Zawadzka, A. (2020). In search of lightweight deployable tensegrity columns. *Applied Sciences, 10*(23). ISSN 2076-3417.

Chapter 7
Biotensegrity: Concept, Principles and Applications

Graham Scarr

Abstract Biotensegrity models living systems in ways that were inconceivable in the past but has taken some time to become widely accepted because of its challenges to generally accepted wisdom. Orthodox biomechanics has its origins in mechanistic models from the seventeenth century and allowed over-simplified representations of anatomy and motion to persist to the present day, with the approximations and assumptions inherent within its methods routinely overlooked. In contrast, a biotensegrity perspective recognizes that the human body is NOT a machine but a complex heterarchical structure of intertwined relationships that result from a fundamental set of self-organizing principles. Here, the closed-chain kinematics of tensegrity provide a unified mechanical system that extends from molecules to the complete organism and represents a paradigm shift in thinking. A new classification of biomechanical models that includes super-stable tensegrity and has relevance to mechanical engineering in general is then introduced.

Keywords Closed kinematic chains · Energy-efficiency · Platonic solids · Self-organization · Super-stability · Systems biology

7.1 Introduction

In 1944, Erwin Schrödinger asked the question "What is life?" and examined it in relation to the scientific knowledge available to him at the time, and many others have done the same. In the process, Schrödinger recognized the shortcomings in conventional biophysical explanations and suggested that "we must be prepared to find a new type of physical law" (Schrödinger, 1944), but after eighty years, how much closer are we to answering that question.

Biology, as the scientific study of life, is a field of research that has long intrigued the human mind, and it is huge. Not only does it examine the form and development of an immense variety of different organisms, it also provides insights into their

G. Scarr (✉)
Ezekiel Biomechanics Group, Nottingham, UK
e-mail: gscarr3@gmail.com

F. Fraternali and J. J. Rimoli (eds.), *Tensegrity Systems*, CISM International Centre for Mechanical Sciences 2, https://doi.org/10.1007/978-3-031-82283-4_7

complex anatomies, mechanics and relationships at every level: from molecules and cells to adult populations and complex ecologies. Our understanding of living systems thus forms a body of knowledge that is unrivalled in any other sphere, and naturally a significant part of that research has been oriented towards improving the human condition and benefitted our healthcare systems in remarkable ways.

A deeper understanding of the mechanics of molecules, cells, tissues and their environments has led to developments in pharmacology that benefit us all, from simple pain killers to the treatment of life-threatening diseases. In-silico models provide valuable information that informs the development of new types of prosthesis and super-complex surgeries, while the involvement of tensegrity in the behaviour of cells, formation of cancers and the body as a whole has long been appreciated (Kirkby, 1975; Robbie, 1977; Ingber et al., 1981; Levin, 1981).

However, current thinking about the human body is largely based on mechanistic models from the seventeenth century (Borelli, 1680; Piolanti et al., 2018) and allowed over-simplified representations of anatomy to persist to the present day. Even though the dissections of Vesalius (1543) were truly revolutionary in the sixteenth century, established traditions and the persistence of outdated methodologies have unwittingly constrained how we think about life (Levin & Lowell de Solórzano, 2024; Nicholson, 2018; Scarr et al., 2024).

Living organisms are hugely complex, fundamentally indeterminate and exist in states that are far from equilibrium (Glazier & Davids, 2009; Pascal & Pross, 2015), and as a consequence, they have been simplified in ways that have enabled great progress in the mapping of structure to function but obfuscated the foundational basis for stability, motion and life itself (Profeta et al., 2020; Scarr et al., 2024). As Russell (2009, p. 42) wryly noted, “All exact science is dominated by the idea of approximation”, which suggests that biomechanical models have yet to reach their full potential because much of the basic knowledge that currently underpins them is questionable.

Biotensegrity is a concept that challenges many traditional ideas because it recognizes that all living systems conform to some fundamental principles of self-organization and that these apply at every level from molecules to the complete organism (Kiely & Collins, 2016; Levin, 2015; Scarr, 2018). However, trying to explain its significance is not a straightforward process. The human body is NOT a machine, and like the tensegrity model (Fig. 7.1), a reductionist examination of its constituent parts in isolation does not tell us much about the emergent properties of the whole (Levin & Lowell de Solórzano, 2024). As that old adage says: ‘Before embarking on a new field of study, make sure you understand it thoroughly first’, which implies that it is necessary to have an appreciation of both the details *and* the complex picture at the same time; and as we have to start somewhere, this chapter examines the subject from several different perspectives. Certain aspects might then seem rather simplistic on their own but must always be considered in relation to the whole. As Levin put it: “Tensegrity is not an engineering principle that applies to biology, but a biological principle that can be applied to engineering” (Pers. comm. 2024).

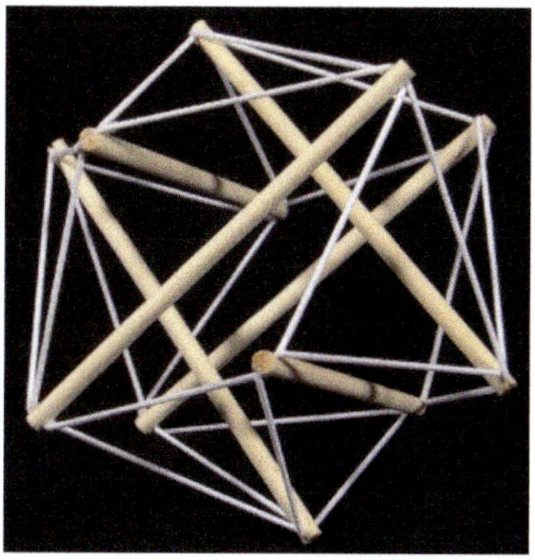

Fig. 7.1 6-strut tensegrity (T-icosahedron, T-icosa, expanded-octahedron); © Rory James

Biotensegrity represents a completely different way of thinking about the human body—and all living systems—where stability and motion are controlled from within the structure itself and in synergy with the nervous system (where befits) (Kiely & Collins, 2016; Levin et al. 2017; Wilson & Kiely, 2016). The strength of individual 'parts' is then no longer the primary concern—as it is with engineering models—but rather the dynamic balance of invisible force vectors within them (Ingber et al., 2014; Kiely & Collins, 2016; Stamenović & Smith, 2020). It is thus important to note that the terms *tensegrity* and *biotensegrity* can mean different things that depend on the context, and the following will be a qualitative assessment of the concept—principles and mechanics—and its relevance to engineering in general.

7.2 Employed Notation

This chapter considers *discontinuous compression* tensegrity structures as assemblies of *cables* that carry tensile forces, 'floating' *struts* that carry compressive forces, and *bars* that carry either tensile *or* compressive forces.

7.3 How We Got to Where We Are Now

Foundational to the modelling of human movement has been the work of Borelli (1680) who compared living organisms with man-made machines of the day and described the body as a system of levers, pulleys and other contrivances that are now part of mainstream thinking (Humphrey, 2003; Piolanti et al., 2018). In other words, a musculoskeletal duality of weight supporting columns, beams and arches with the bones stacked on top of one another like a pile of bricks and the soft-tissues moving them around like the rigging in a sailing ship, and such a system of rigid and stretchable elements that mechanically complement each other is naturally very appealing and computationally well-supported, but it also creates a problem (Levin, 2002; Scarr et al., 2024).

The stress/strain simplification. Current biomechanical theory is largely founded on the laws of classical mechanics as formulated by Galilei, Newton and Hooke, etc., but these 'rules' were based on the behaviour of inanimate objects and do not necessarily apply to the nonlinear dynamics of living tissues (Levin & Lowell de Solórzano, 2024). Hooke's Law, for example, describes the linear relationship between stress and strain with Young's Modulus defining the slope of the curve, but this creates a problem when applied to biology (Fig. 7.2). Living systems typically display a non-linear stress/strain relationship, but because this complicates the analysis, the tissues are generally considered to function within the more 'vertical' high stress/strain region, which is then assumed to be a straight line. Living tissues, however, tend to operate within the more 'horizontal' low stress/strain region, where they will be more responsive, consume less energy and operate more efficiently—all defining characteristics of living systems. As Harris (2018) noted: "to regard Young's Modulus as a constant… is a very big simplifying assumption" because it will be different at each point along the non-linear curve.

Compressed joints. In addition, a characteristic of lever systems is that they generate compressive forces between their joint surfaces, yet the thickness of articular cartilage tends to *reduce* in larger vertebrate species (Malda et al., 2013), while a clear 'space' has been demonstrated in the human knee (femur-tibia) (Hakkak et al., 2015; Levin & Madden, 2005; Terayama et al., 1980) and elbow (humerus-radius) (Kim et al., 2002) in vivo, even when the bones are pushed together. The ability of cartilage to sustain high compressive loads in the knee and temporo-mandibular (jaw) joints has also been questioned (Andrews et al., 2011; Juran et al., 2013; Kim et al., 2012), while a review of the literature on temporo-mandibular mechanics noted a conspicuous

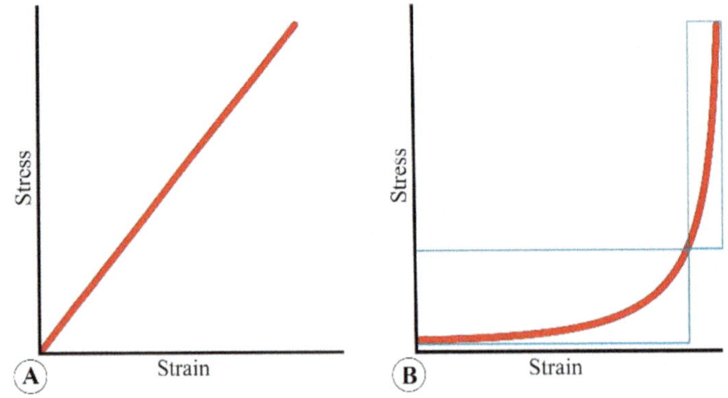

Fig. 7.2 Typical stress–strain curves for materials that conform to **a** Hooke's Law; and **b** the behaviour of living tissues and tensegrity (see text)

lack of in vivo information available to support the lever model (Scarr & Harrison, 2016). To put it bluntly, biomechanics does not have a satisfactory model to explain how synovial joint surfaces could be routinely decompressed, or even ‘pulled apart’, and it is considered that this inadequacy has severely restricted our understanding of normal function.

Transient stress concentrations. “A lever system cannot passively stabilize itself in the presence of unexpected forces and moments, and the assignment of this role to the action of responsive muscles only transfers this latency elsewhere, and ultimately to an external neural controller that is presumed capable of providing the necessary ‘catch-up’ stability. However, even the fastest mono-synaptic reflex experiences a time-delay (Del Vecchio, 2023) which means that the necessary timing and precision required to neutralize the effects of sudden perturbations in the system is open to compromise (Cabe, 2019; Fultot et al., 2019). An abrupt change in force dynamics could create transient peak stress concentrations before the system has had time to react (Hayati et al., 2020; Lin et al., 2019), and although engineered structures can be heavily reinforced against the potentially damaging effects of these, it is likely that this inherent vulnerability would adversely affect living tissues” (Scarr et al., 2024).

“Traditional musculoskeletal analysis has tended to focus on steady-state forces and kinematics and paid little attention to transients, but the transient properties of force generation are immensely important (Lin et al., 2019). Until recently, the rate of change of force over time had no standard term in physics, and as a result, the dynamics of force transients have been simplified through the same analytic methods and formalisms as used in engineering (Deriglazov, 2017). Here, the concepts of force, velocity, momentum, torque and acceleration, etc., are typically processed through 1st and 2nd order differential equations without the need to account for their higher-order derivatives, but the assumption that this approximation is also valid for living tissues is not necessarily the case” (Scarr et al., 2024).

“Put simply, a force impact is not an instantaneous event but a transient one: a derivative of momentum that rapidly escalates and takes time to reach its peak (Fig. 7.3). The transient after acceleration is jerk and followed by snap and crackle, respectively (Eager et al, 2016; Hayati et al., 2020). The 1st derivative of force over time—yank—is then highly significant because it is proportional to jerk (Lin et al., 2019), and an unexpected force impact that generates peak stress concentrations faster than the tissues can react to carries the potential for injury at relatively low speeds and forces” (Scarr et al., 2024).

The effects of these mechanical transients coupled with the long healing times required for post-traumatic recovery, then suggest that such peak loadings and intractable instabilities (inherent in the lever model) would lead to material fatigue and destructive tissue failure, which further highlights the problem with orthodox biomechanical thinking. However, before examining the advantages of tensegrity mechanics to living systems, it is appropriate to introduce some of the key players involved.

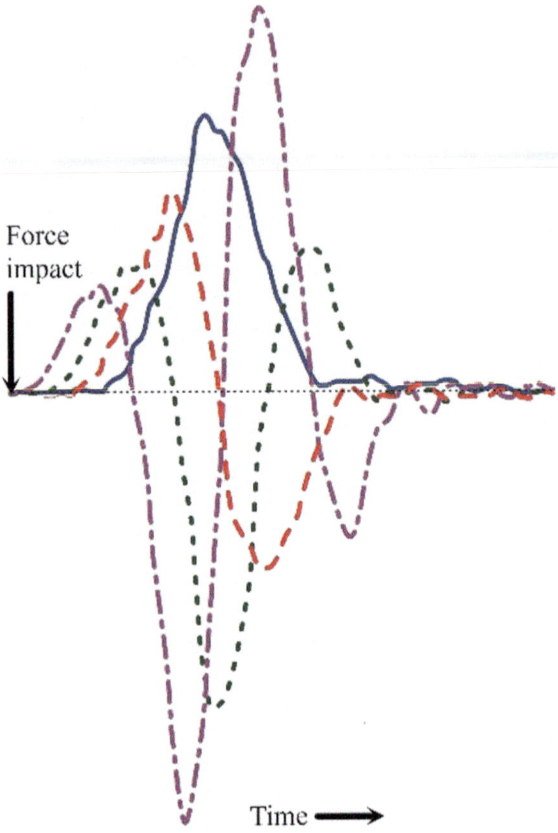

Fig. 7.3 An illustration of transient dynamics that follow a force impact. Acceleration (blue, solid), Jerk (red, dashed), Snap (green, dotted) and Crackle (magenta, dash-dotted). Adapted from Eager et al. (2016) under Creative Commons Attribution 3.0 licence

7.4 The Origins of Biotensegrity

R. Buckminster Fuller. As a maverick thinker and inventor, Fuller (1895–1983) had been developing a system of geometry based on first principles since the 1920s, and not satisfied with conventional explanations that used the cube with its 90° angles as the frame of reference, considered that a coordinate system based on 60° was far more fundamental, and developed his theory of *Synergetic*s as 'the study of nature's coordinate system' (Fuller, 1982; Edmondson, 2007). Fuller perceived that all natural forms were the result of matter being acted upon by invisible forces and proposed them as finite energy systems consisting of tension and compression acting together. He also recognized the sphere as the ultimate compression element—because it resists external forces equally from every direction—and focused on a shape that mimics it structurally: the icosahedron with its twenty triangular faces that can be subdivided and form higher-order icosahedra that come ever closer to approximating the sphere. Fuller is thus best known for his geodesic domes (Krause and Lichtenstein, 1999), but it is the icosahedral symmetry that is most significant here.

The geometry of the geodesic dome was the inspiration that resolved the outer structure of the spherical viruses and fullerene molecule C60 (Caspar, 1980; Kroto, 1990), and also appears in radiolaria, pollen grains and clathrin vesicles. Fuller also recognized the bicycle wheel as a tensegrity structure, with the compressed hub and outer rim held both apart and together by the tensioned spokes—and which maintain the wheel's integrity—and the mechanics of the shoulder (Levin, 1997), pelvis (Levin, 2007) and temporo-mandibular joints (Scarr & Harrison, 2017) have all been compared with the bicycle wheel. Fuller was thus using his new method of geometry to demonstrate a dynamic architecture based on nature's principles and began spreading the word.

Kenneth Snelson. As a young art student attending Fuller's lectures in 1948, Snelson (1927–2016) was so impressed that he returned the following year with his own sculpture (the *X-Piece*) constructed from pieces of wood that were suspended—or appeared to float—within a tensioned network of nylon cables (Snelson, 1996), and Fuller immediately recognized its significance. "Simple as this pioneering work was, it pointed ahead to the possibility of structures in which form and function truly are... one, and the visible configuration of the sculpture is simply the revelation of otherwise invisible forces" (Heartney, 2009, p. 20); and this statement is the key to appreciating tensegrity in a biological context: *the physical representation of invisible forces*.

Both Fuller and Snelson contributed to the development of tensegrity but ultimately went their separate ways. By the 1970s, Fuller was becoming a household name with his geodesic domes springing up all over the place (Krause and Lichtenstein, 1999), but he always maintained that tensegrity was really a design principle of nature (Fuller, 1982). Snelson successfully pursued a career in sculpture but remain convinced that tensegrity had no real value except in art (Snelson, 1996), yet the irony of this is that his sculptures have been the inspiration to others in exploring tensegrity in a biological context.

Donald E. Ingber. As an undergraduate student in 1975, Ingber recognized similarities between a tensegrity model and the behaviour of cells when compressed—with both springing back to their original shape when released—a finding that later inspired his research as a cell biologist and work on the architecture of cancerous tissues (Ingber, 2023). Since then, Ingber's work has provided much of the empirical support for tensegrity in integrating the mechanics of the cellular cytoskeleton and extracellular environment with internal chemical signalling cascades, changes in cell function and the development of complex organs and tissues (Ingber et al., 2014).

Stephen M. Levin. As an orthopaedic surgeon in the 1970s, Levin noted things at the operating table that could not be explained by conventional mechanical theory and began searching for more plausible explanations by examining the skeletons of dinosaurs and other animals (Levin, 2002, 2015). Dinosaurs are essentially no different from any other vertebrate except that they could weigh up to 100 tons, and as a doubling in height implies a cubic increase in volume and mass, Levin wondered how the immense weight of these creatures could be supported and moved through a lever system that would have placed enormous stresses on their tissues, especially as the neck and tail can extend up to 10 m from their bodies.

He also noticed that while most engineered structures transfer their weight to the ground through continuous compression, the struts within Snelson's tensegrity structures are suspended within the tensioned network of cables and form a truss system that distributes mechanical stresses through the system in a way that remains completely stable, irrespective of the direction of gravity. Snelson's sculptures then stimulated the idea that tensegrity might provide a viable alternative to traditional explanations because it can be applied at every structural level within a complex heterarchy and avoids potentially damaging stress concentrations (Levin, 1982, 2002, 2015) During development, tension and compressional forces would interact in ways that lead to the formation of complex tissues and conform to a fundamental set of self-organizing principles.

7.5 Principles of Self-organization

At the most basic level, atoms interact with each other in several different ways—van der Waal's forces, electrostatics, hydrogen bonding, covalency and steric-repulsion etc.—all of which can be categorized into those that 'pull' atoms together and others that keep their centres apart. Of course, there is a lot more detail associated with these processes, but simplifying then into this duality of attraction and repulsion reveals them to be the drivers that cause atoms to spontaneously organize themselves into the most stable and energy-efficient configurations as mineral crystals and more flexible molecules (Levin, 2015; Scarr, 2018).

Such arrangements form as they do because of some fundamental principles of self-organization: geodesic geometry, close-packing and the minimization of energy—and which apply at *every* size-scale—where *attraction and repulsion* at higher size-scales are now referred to as *tension and compression* and can be modelled through tensegrity. Crystals and molecules then become close-packed tensegrity configurations of atoms with invisible struts (Bywater, 2017; Edwards et al., 2012; Eichenauer & Lordick, 2019; Reilly & Ingber, 2018).

Tensegrity frameworks are inherently geodesics because their structural members transfer tension and compressional forces along the shortest pathways and in straight lines. They are close-packed because their component parts *automatically* get as close as they can to each other, and they *always* settle into a state of minimal-energy. The tension elements pull the nodes together, and the compression elements keep them apart, with the tensegrity framework being the embodiment of these three basic principles and the *physical representation of the invisible forces within it*. The intrinsic tension (self-stress) contained in both living tissues and tensegrity models also means that the initial part of the stress–strain curve never reaches zero stress and that they are always 'primed for action' (Masi & Hannon, 2009) (Fig. 2b).

So, rather than thinking of the human body in terms of separate 'bits', like a machine, a biotensegrity perspective considers their relationships within a complex heterarchical system and examines each 'part' in relation to the whole (Levin, 2015; Scarr, 2018). We might then think that *tensegrity* represents the way forward in biomechanics, but there is a potential problem with this word in a biological context because of the way it has been applied in mechanical engineering.

Tensegrity or biotensegrity. Both Fuller (1982) and Snelson (1996) emphasized that the compression members within a tensegrity framework remain distinct—"continuous tension and discontinuous compression"—while Skelton et al. (2001) relaxed this requirement and defined a class-k tensegrity structure as "a stable equilibrium of axially-loaded elements with a maximum of k compressive members connected at the node". Here, a node (joint) can include any number of compression members with just one defining class k-1, two defining class k-2, and so on, which means that Fuller and Snelson's definitions only apply to class k-1. So, while this classification is useful in an engineering context, the observations that living bones are not heavily compressed—and can even maintain a space between their articular surfaces (Andrews et al., 2011; Hakkak et al., 2015; Levin & Madden, 2005; Terayama et al., 1980)—implies that Skelton et al's (2001) definition is not applicable to living systems, and as a consequence, Levin (2002) introduced the term *biotensegrity* in order to make this distinction clear.

Of course, both fields are closely linked and can benefit from each other, but while the term *tensegrity* refers to the mechano-structural principle, the concept of *biotensegrity* relates more to the invisible force vectors and dynamic complexity of the system, i.e. "a world of processes rather than a world of things" (Dupre & Nicholson, 2018, p. 4). So, although most of the bio-focused literature has used the word tensegrity out of tradition (Boghdady et al., 2021; Ingber, 2023; Turvey & Fonseca, 2014), it should be noted that the two terms are *not necessarily* synonymous. Living organisms are relatively light, flexible and capable of functioning much the same in any position, and because they have refined themselves over hundreds of millions of years of evolution, it is likely that they have already found the most energy-efficient ways of doing things.

7.6 The Mechanics of Life

Closed-chain kinematics. Closed kinematic chains (CKCs), or linkage systems, have long been employed in mechanical engineering because they provide a simple and efficient mechanism with multiple applications. Here, multiple bars are coupled into continuous mechanical loops with each one influencing the position, motion, and stability of all the others and enabling the controlled transfer and amplification (or attenuation) of force, speed and kinetic energy (Fig. 7.4). CKCs have also been described in the feeding mechanisms of fish (Farina et al., 2019; Muller, 1996), shrimps and reptiles; and the jaws and limbs of humans (Burgess, 2021; Levin et al., 2017; Olsen, 2019; Scarr & Harrison, 2017; Valero-Cuevas et al., 2007).

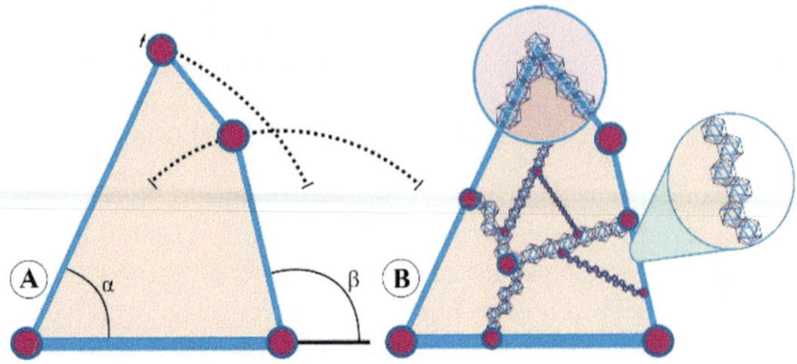

Fig. 7.4 **a** 4-bar mechanism showing the 'pin-joints', trajectories and limits of three moving bars in relation to a reference bar, and the changing angles that define its mechanical properties., adapted from Levin et al. (2017) © Elsevier; **b** coupled 3/4-bar configurations within a tensegrity hierarchy., adapted from Scarr (2025) © Handspring

Kinematics is all about the geometry of motion and generally described through algebraic equations, yet a graphic portrayal of their geometries reveals things that might otherwise not be appreciated, and which are particularly relevant here because closed-chain kinematics constitutes the basic mechanics of tensegrity (Fig. 7.5). The 4-bar mechanism is then highlighted because it enables the structure itself to control the position and motion of every part of the system—from molecules to the complete organism—with the 3-bars creating the stability that allows the 4-bars to change shape, while those with five or more bars are uncontrollable on their own but can be part of more complex systems (Rimoli & Pal, 2017).

The controlled transfer of power. Of course, these CKCs might just appear to be complex lever systems, but this is not the case from a bio-perspective where each anatomical 'bar' and 'node/pin-joint' consists of multiple sub-structures organized in

Fig. 7.5 The 6-strut tensegrity highlighting some of the stable 3-bars (white) and flexible 4-bars (blue dotted) that regulate changes in shape (auxetic contraction) when compressed

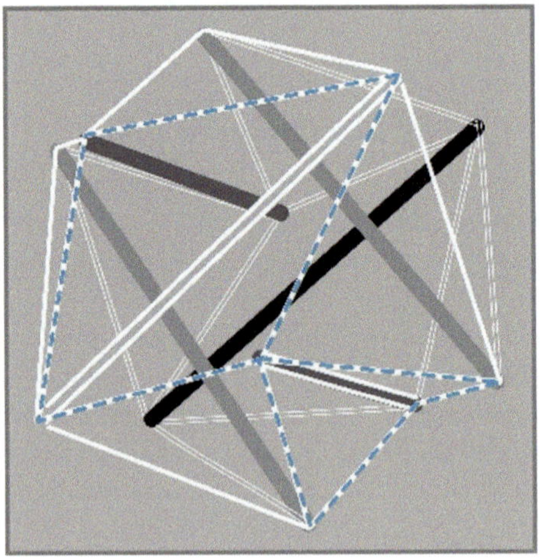

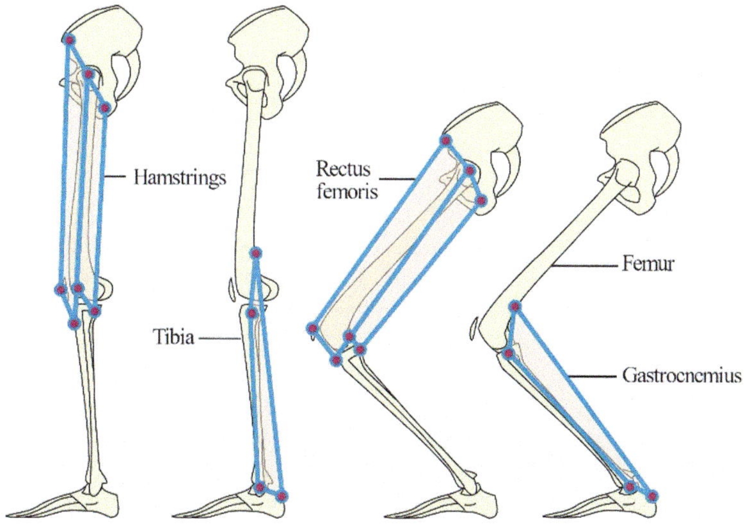

Fig. 7.6 Schematic showing how movement can be controlled by the closed-chain geometries of muscles and bones. Reproduced from Scarr (2025) © Handspring

the same way within a vast global heterarchy that extends down to the molecular level (Reilly & Ingber, 2018) (Fig. 4b). Here, each bar is under tension *or* compression and mechanically integrated with all the others at every level, so that when one part changes in length or position the entire system becomes involved. Mono-articular muscles can then be considered as variable-length bars within 3-bar systems, while bi-articular muscles can form tensionally stiff bars within 4-bar systems that enable multiple joints to be coupled together, and in ways that enable their power to be transferred from one part of the body to another in a controlled way (Dischiavi et al., 2018; Levin et al., 2017) (Fig. 7.6).

Such closed-chain systems are then significant because they enable motion to be controlled from within the structure itself and in ways that are beyond the sole capability of the nervous system (Saxena et al., 2012; Valero-Cuevas et al., 2007). They also enable the tissues to respond *instantly* to rapidly changing conditions and operate much faster than muscles can contract—thus overcoming the problems with the lever model, force transients and potentially-damaging stress concentrations (Sect. 7.3)—and allowing muscle generated forces to be transmitted over a much wider field than is generally appreciated (Dischiavi et al., 2018; Levin & Scarr, 2022) with degeneracy built into the complexity of the system (Wilson & Kiely, 2016).

Evolutionary implications. Muller (1996) reviewed the mechanics of the 4-bar mechanism in animals and highlighted two important features. First, that changes in shape map *non-linearly* to the kinematic transmission ratio (KT)—an emergent property that also *characterizes the behaviour of living tissues* (Fig. 2b); and second, that changes in the relative lengths of the bars leads to a redundant *many-to-one* *mapping* (MTOM), where different 4-bar shapes can produce the same mechanical output (KT), and with evolutionary implications because it permits a partial decoupling between morphological diversity and function. In other words, it explains how one part of the anatomical CKC (tensegrity) system could change in shape—whilst keeping the same mechanical functions—within a global collective of gradual step-by-step processes that ultimately leads to the transition of one species into another (Alfaro et al., 2005; Muñoz, 2019). It is also likely that such MTOM is involved in embryological development as each anatomical 'part' gradually evolves in shape, position and function in response to the mechanics of its local environment, changes in cellular behaviour (Ingber, 2023) and those basic principles of self-organization (Sect. 7.5).

Living organisms are thus not static but in a constant state of flux (Nicholson, 2018; Pascal & Pross, 2015), and all the examples in this chapter are snapshots of systems that are constantly changing. The planar 4-bar is then highlighted because of its mechanical simplicity and as the first iteration of the omni-directional CKC/tensegrity heterarchy (Figs. 7.4 and 7.5), with the behaviour of each part dependent on both its internal structures and the wider environment (Cabe, 2019; Turvey & Fonseca, 2014), and the system as a whole controlling the flow of kinetic energy (Bejan, 2016). It should also be noted that the T-icosa model contracts auxetically when compressed (Fig. 7.5), and such an emergent characteristic that increases density/pressure within the structure has been described in the heart and bladder (Levin, 2002), tendons (Gatt et al., 2015) and other soft-tissues (Ziauddin & Ali, 2013). The distinctive mechanical behaviours of biological tissues then makes a case for a distinction to be made between different types of model.

7.7 Classifying Biomechanical Models

Humphrey (2003) stated that: "the overall goal of biomechanics is… the general improvement of the human condition"—and which is a noble cause that no-one would wish to deny—but *differences* between the models used to fulfill that goal are often poorly appreciated. "While medical and social needs drive the enormous amount of research that takes place at every level, the employment of reductionism and engineering methodologies as a way of facilitating its goals has not been entirely consistent with the unbiased progression of anatomical knowledge (Dischiavi et al., 2018; Nicholson, 2019). Indeed, the ability to overlook the assumptions, approximations and unknowns inherent within its methods is routinely accepted as part of the modelling process, with such matters considered as minor inconveniences that can be safely ignored or re-examined following the acquisition of more data" (Scarr et al., 2024); and with this in mind, a new classification of biomechanical models that places them in a functional context is now introduced.

Class 1: The observable reality of living systems, but which are fundamentally indeterminate (Glazier & Davids, 2009).

Class 2: Interpretations that support a particular hypothesis or viewpoint and appear to emulate Class 1, e.g. most bio-oriented research, levers and CKCs.

Class 3: Engineered solution-oriented mechanisms that fill some medical or social need but clearly differ from Class 1, e.g. prosthetics, robotics and biomimicry.

Class 4: Engineered structures inspired by Class 1 but do not pretend to be representations of biology, e.g. machines and architecture (Gomez-Jauregui et al., 2023).

Sub-classes of 3 and 4: Engineered structures with relevance to class 2 modelling from a biotensegrity perspective, e.g. geodesic dome, bicycle wheel and Snelson's sculptures.

Here, the accurate modelling of living systems (Class 1) must be considered aspirational because Class 2 interpretations of reality are always likely to be incomplete (although *biotensegrity* now presents a conceptual modelling framework (see Sect. 7.10)). Class 2 models fill the contents of most textbooks and research papers—*and are essential to the progress of human knowledge*—but can become established within a tradition that erroneously accepts them as accurate representations of reality (e.g. levers, machines). Classes 3 and 4 are then substantially different from Class 1, although greatly influenced by it, because they are oriented towards practical solutions and can use whatever means necessary to achieving this. Unfortunately, many of the simplifying assumptions that are legitimately made in Classes 3 and 4 have been incorporated into Class 2 models with embedded historical dogmas, outdated methodologies and the best of intentions creating a subjective bias simply because there seemed to be no viable alternative.

Of course, there is a fair amount of overlap and flexibility between these classes, and which are largely presented here as 'food for thought' with the distinctions between them being most significant. Here, Classes 3 and 4 are constrained by the properties of the materials, practicality of construction and cost, while living systems (Class 1) are constrained only by those fundamental principles (Sect. 7.5).

Class 2 tensegrity models have been described in relation to molecules (Azulay et al., 2022; Bywater, 2017; Edwards et al., 2012; Reilly & Ingber, 2018), viruses (Sitharam & Agbandje-Mckenna, 2006), spider silk (Fraternali et al., 2020), cells (Ingber et al., 2014; Fraldi et al., 2019), tissues (Boghdady et al., 2021; Ingber, 2023) and regional anatomy, including the cranial vault (Scarr, 2008), temporo-mandibular joint (Scarr & Harrison, 2017), spine (Levin, 2002), shoulder (Levin, 1997), elbow (Scarr, 2012), pelvis (Levin, 2007; Pardehshenas et al., 2014), knee (Kim & Turvey, 2018), foot (Kiely & Collins, 2016; Wilson & Kiely, 2016), heart (Parker & Ingber, 2007), avian lung (Maina, 2017), human lung (Ingber, 2023), brain (Van Essen, 2023), postural control (Cabe, 2019; Doyon et al., 2019; Mangalam et al., 2020), homeostasis (Stamenović & Smith, 2020) and the body as a whole (Dischiavi et al., 2018; Levin, 2015; Levin & Lowell de Solórzano, 2024; Scarr, 2018).

Class 3 models of the spine (Melnyk & Pitti, 2018; Oh et al., 2022) and limbs (Jung et al., 2021; Brandão and Savi, 2022; Abourachid et al., 2023) have also been developed in relation to prosthetics and robotics, but it is important to note that Class

2 tensegrity models *cannot* be meaningfully assessed from an engineering (Class 3/4) perspective (Gomez-Jauregui et al., 2023) without considering their biological (Class 1) complexities.

The role of tensegrity has also been described in plant physiology (Mirabet et al., 2011; Witztum & Wayne, 2014), and recognized within cancer research (Dinicola et al., 2011; Tadeo et al., 2014), clinical decision-making (Crowle & Harley, 2020; Fullerton, 2018; Scarr, 2020), social systems (Primus, 2022) and team sports (Caldeira et al., 2020), illustrating the scope of this relatively new systems-biology modelling approach. Understanding the benefits of a biotensegrity perspective to biomechanical modelling, prosthetics, robotics and mechanical engineering in general thus means starting at the beginning.

7.8 A New Type of Physical Law

The Italian Renaissance witnessed a dramatic increase in knowledge that now forms the basis for modern scientific thinking, but there is a problem. The reductive methods of science have tended to concentrate their efforts on the genetic code, molecular pathways and structural analysis, and largely ignored the simplicity of those fundamental rules of self-organization because they are 'just too simple' (Sect. 7.5). So, while Fuller's theory of Synergetics as *the study of nature's coordinate system* can seem rather cryptic and should not be taken as explicitly scientific, it contains ideas that are revolutionary in their scope and depth (Fuller, 1982; Edmondson, 2007).

The Platonic 'Solids'. Geometric patterns and shapes do not just appear out of nowhere but because of some basic physical rules (Bejan, 2016; Edmondson, 2007), and Fuller recognized the significance of the Platonic 'solids' as a set of regular shapes that emerge from a sphere: the centrepiece of his universal system of geometry (Fig. 7.7). Here, their edges can be projected onto the surface of a sphere and divide it into faces of equal size, while multiple spheres can close-pack into hexagonal arrangements in 2-D and the Platonic shapes in 3-D (Eichenauer & Lordick, 2019).

Bubbles and foams. Bubbles spontaneously form spheres because the pressures acting on their surfaces are balanced in every direction, but true spheres don't exist in the real world because they are just mathematical constructs. Soap bubbles may look like spheres but have surfaces that consist of lots of irregularly shaped molecules that maintain the overall shape through surface tension. Fuller then appreciated that while a high-order icosahedron closely approximates the space-filling efficiency of the sphere, this can be represented by the T-icosa model as the omnidirectional balance of tension and compressional force vectors (Fig. 7.1) and the archetypal Class 2 model for biology (Levin, 2002).

Bubbles naturally follow Plateau's Laws and spontaneously organize into arrays that minimize the overall surface area, with their surfaces always meeting in groups of three, and edges that always intersect in fours at the *tetrahedral angle* of 109.47°

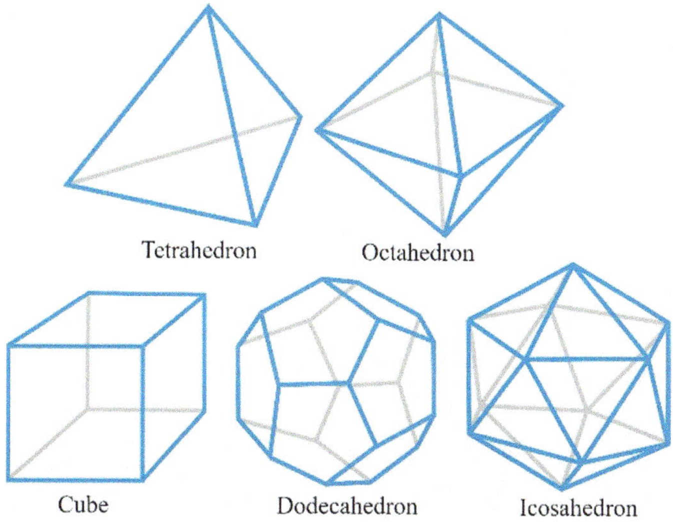

Fig. 7.7 The five Platonic 'Solids'

(Serón et al., 2022). Similar-size bubbles automatically form hexagonal arrangements, and this pattern also appears in the close-packing of actin and myosin molecules within muscles (Standring, 2005, p. 118), parabronchi within the avian lung (Maina, 2017), eyes of insects and honeycomb of the bee, etc. (Thompson, 1917). Living cells frequently close-pack into hexagonal patterns but are not constrained solely by Plateau's Laws because of the higher levels of complexity (Graner & Riveline, 2017). The tetrahedral angle also appears at the junctional offset between hexagonal cells on either side of the bee's honeycomb, and closely matches the angular connections between trabeculae within bone (Reznikov et al., 2016).

Helices and crossed-helical tubes. The tetrahedron is then further interesting because it can close-pack with others to form a *tetrahelix* (Boerdijk–Coxeter helix) (Lord & Ranganathan, 2001) as the geometric basis for molecular helices (Lord, 2002; Sadoc & Rivier, 2000), with the helix being a common motif in protein structures and a general model for coiled winding in a wide variety of different organisms and tissues at many different levels (Scarr, 2016); and ultimately, crossed-helical arrangements of collagen fibres that reinforce the tubular walls of arteries, muscles, gastrointestinal tract and intervertebral discs and mechanically control changes in their shape (Ellers et al., 2024; Scarr, 2016). In addition, the faces of the pentagonal dodecahedron give rise to the Golden Ratio, Fibonacci Sequence and helical/spiral configurations in plants (Rivier et al., 2016), cephalopods, mammals (Scarr, 2016) and humans (Persaud-Sharma & O'Leary, 2015).

The geometries of the Platonic 'Solids' have thus been linked with the close-packing efficiencies of spheres (icosahedron) and helices/spirals (tetrahedron/dodecahedron), and directly relate to those basic principles of self-organization and tensegrity (Sect. 7.5), with (bubble) foams being the simplest examples of soft-matter physics (Drenckhan et al., 2005; Rivier et al., 2016) that also includes the complex behaviours of polymers (molecular helices), colloids, surfactants and liquid crystals within living systems (Hirst, 2013; van der Gucht, 2018). Simplicity and complexity are thus more closely related than is often appreciated with all the above now addressing Schrödinger's (1944, p. 80) suggestion that "we must be prepared to find a new type of physical law", and the concept of biotensegrity changing traditional thinking about anatomy and the mechanics of life.

7.9 Updating Anatomy

The notion that the musculo-skeletal duality constitutes the basic structural framework of the body is now well-established but was conceptually rivalled in the eighteenth century by a system of connective tissue *compartments* that Lamarck (1963, p. 230) declared to be "…the universal matrix of all organization", but gradually faded in significance as the 'muscles moving bones/lever paradigm' assumed functional dominance (Scarr et al., 2024) (Sect. 7.3). As a consequence, the fascial tissues have been widely dismissed as mere packing tissue or relegated to ancillary roles but are now recognized for their involvement in structural stability, motion control and the aetiology of cancers and other pathologies (Dischiavi et al., 2018; Wilke et al., 2018).

The term fascia refers to the soft collagen-rich connective tissues that form a complex fibrous/fluid (soft-matter) network that surrounds virtually every cell in the body as the extracellular matrix (ECM) (Mao & Wickstrom, 2024) and is continuous with the tougher fibrous sheaths that *enclose and permeate every organ, muscle and tissue*, so that at the microscopic level there is no clear distinction between where one part ends and another begins (Guimberteau et al., 2024).

A reassessment of those forgotten tissue compartments from a biotensegrity perspective (Scarr et al., 2024) recently considered the fascial system as a globally-tensioned fibrous network encompassing a complex heterarchy of regionally specialized compartments under compression (i.e. tensegrity), and concluded that this intrinsically-stable functional entity constitutes the basic structural/architectural framework of the human body. Here, muscles and bones are no-longer separate entities but compartmental specializations of the global fascial system, and which together with the underlying mechanics of CKC/tensegrity suggests that the fascial system should now be considered as the primary controller of movement and acts in synergy with the nervous system (where befits) (Cabe, 2019; Fultot et al., 2019); and which effectively renders the term *musculo-skeletal* obsolete.

The intimate relationship between the cytoskeleton and ECM is then particularly significant because it mechanically couples the latter with internal chemical signalling cascades and changes in cell function (Boghdady et al., 2021; Ingber, 2023) and influences the formation and behaviour of the fascial system and body as a whole through the mechanics of CKC/tensegrity.

7.10 Stability Before Mobility

Super-stability. Traditional biomechanics has always tended to focus on movement with muscles and bones taking centre stage and the nervous system providing the necessary catch-up stability for transient muscular latencies, and some of the problems with this were described in Sect. 7.3. The reassessment of the fascial system as an intrinsically-stable architecture of tensioned fibrous tissues and compressed compartments now offers an alternative approach (through biotensegrity), but an issue now arises with the word *stability* because it evokes images of firmness, fixity and permanency while stability in dynamic systems is inseparable from its opposite: change (Bejan, 2016; Pascal & Pross, 2015). Textbook descriptions of anatomy can give impressions of unchanging stability but are really just Class 2 snapshots of dynamic systems that are constantly changing.

So, if this intrinsically stable framework is constantly changing, there must be some sort of homeostatic algorithm that keeps all its internal processes orderly (Stamenović & Smith, 2020), but until recently there has been no ready-made math for determining whether such a framework really is intrinsically stable, which is why biomechanics has been dominated by heuristic approaches that can at least provide partial solutions (Freutel et al., 2014). Connelly and Back (1998), however compared the mathematics of tensegrity with the spider web and highlighted the algorithmic framework of super-stability—a fundamental theorem that now formalizes biotensegrity as a viable model for living systems (Fig. 7.8).

Super-stability encompasses the tensegrity variables of tension and compression and has relevance to living systems because they conform to the same self-organizing principles. Here a system is mathematically represented as a set of points (vertices, nodes) connected by edges with simple distance constraints, with certain combinations that can get closer to each other but not further apart (cf. tension) and others that can get further apart but not closer together (cf. compression). The edges are thus representing tension and compression springs that can change in length and vary the amount of stored elastic-energy (Connelly & Guest, 2022; Yang et al., 2019).

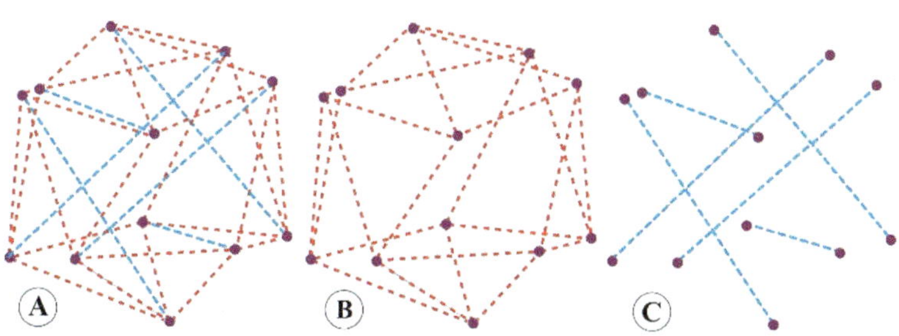

Fig. 7.8 Super-stability: **a** Tensegrity (T-icosa) portrayed as an energy framework and set of points (vertices, nodes); **b** point combinations (edges) that can get closer together but not further apart (cf. tension); and **c** point combinations (edges) that can get further apart but not closer together (cf. compression). Adapted from Levin and Scarr (2022) © Elsevier

The stability of the system is then achieved through coupled changes in the elastic strain energies as the springs automatically seek their minimal states, and because these energy functions are split between positive (tension) and negative (compression), such a system will always resolve itself into the most stable and energy-efficient configuration at any given moment in time; but as a super-stable framework, it can have ANY number of possible edge combinations for ANY number of vertices and without restricting the assignments.

Here, the edges can be freely split into shorter ones with new vertices, thus increasing their total count within the framework, or connected into longer ones that decrease the number of vertices, and all without compromising the system's stability. In other words, the edges can get longer, shorter, divide, merge or even switch energy functions with the entire system involved in making these transitions, and the configuration as a whole remaining completely stable throughout—no matter how complicated it becomes—and the value of this as a representation of dynamic anatomy cannot be overstated.

At the highest, most abstract level, super-stable tensegrity provides the intrinsic-stability solution for ANY comparable framework of edges and vertices, regardless of the number of elements within it, and it does so through the minimal-energy principle. Super-stability is essentially a representation of infinity, and the imagery of such a framework that can freely switch between an infinite number of different combinations of edges and vertices then allows a shift to be made from the virtual, abstract and logical perspective to the more tangible morphological transformations of biology and its processes (Dupre & Nicholson, 2018). Here, the local, regional and global meta-stabilities of attraction/repulsion and tension/compression are constantly re-organizing themselves in response to the ever-changing environment and enabling each 'part' to grow, multiply, transform, repair and move in the most energy-efficient ways with the system as a whole maintaining its stability.

In other words, a complex multi-level framework that is constantly changing and maintaining the viability of the system through its myriad interconnected processes. Which then places super-stable tensegrity as the ultimate Class 2 biomechanical model with biotensegrity as the super-stability hypothesis for biology and the conceptual basis from which the new biomechanics can proceed.

References

Abourachid, A., Chevallereau, C., Pelletan, I., & Wenger, P. (2023). An upright life, the postural stability of birds: A tensegrity system. *Journal of the Royal Society Interface, 20*, 20230433.

Alfaro, M. E., Bolnick, D. I., & Wainwright, P. C. (2005). Evolutionary consequences of many-to-one mapping of jaw morphology to mechanics in labrid fishes. *The American Naturalist, 165*, 140–154.

Andrews, S., Shrive, N., & Ronsky, J. (2011). The shocking truth about meniscus. *Journal of Biomechanics, 341*, 34–55.

Azulay, H., Lutaty, A., & Qvit, N. (2022). How similar are proteins and origami? *Biomolecules, 12*, 622.

Bejan, A. (2016). Life and evolution as physics. *Communicative and Integrative Biology, 9*, e1172159.

Boghdady, C. M., Kalashnikov, N., Mok, S., McCaffrey, L., & Moraes, C. (2021). Revisiting tissue tensegrity: Biomaterial-based approaches to measure forces across length scales. *APL Bioengineering, 5*, 041501.

Borelli, G. (1680). *De moto animalium*. Lugduni batavorum: Petrum Vander Aa.

Brandão, A., & Savi, M. A. (2022). Nonlinear mechanics of a smart biotensegrity human foot Prosthesis. *International Journal of Applied Mechanics, 14*, 2150124.

Burgess, S. (2021). A review of linkage mechanisms in animal joints and related bioinspired designs. *Bioinspiration and Biomimetics, 16*, 041001.

Bywater, R. P. (2017). A tensegrity model for hydrogen bond networks in proteins. *Heliyon, 3*, e00307.

Cabe, P. A. (2019). All perception engages the tensegrity-based haptic medium. *Ecological Psychology, 31*, 1–13.

Caldeira, P., Fonseca, S. T., Paulo, A., Infante, J., & Araújo, D. (2020). Linking tensegrity to sports team collective behaviors: Towards the group-tensegrity hypothesis. *Sports Medicine, 6*, 24.

Caspar, D. L. D. (1980). Movement and self-control in protein assemblies. Quasi-equivalence revisited. *Biophysics Journal, 32*, 103–138.

Connelly, R., & Back, A. (1998). Mathematics and tensegrity. *American Scientist, 86*, 142–151.

Connelly, R., & Guest, S. D. (2022). *Frameworks, tensegrities, and symmetry*. Cambridge University Press.

Crowle, A., & Harley, C. (2020). Development of a biotensegrity focused therapy for the treatment of pelvic organ prolapse: A retrospective case series. *Journal of Bodywork and Movement Therapies, 24*, 115–125.

Del Vecchio, A. (2023). Neuromechanics of the rate of force development. *Exercise and Sport Sciences Reviews, 51*, 34–42.

Deriglazov, A. (2017). *Classical mechanics: Hamiltonian and Lagrangian formalism* (2nd ed.). Springer Nature.

Dinicola, S., D'Anselmi, F., Pasqualato, A., Proietti, S., Lisi, E., Cucina, A., & Bizzarri, M. (2011). A systems biology approach to cancer: Fractals, attractors, and nonlinear dynamics. *OMICS A Journal of Integrative Biology, 15*, 93–104.

Dischiavi, S. L., Wright, A. A., Hegedus, E. J., & Bleakley, C. M. (2018). Biotensegrity and myofascial chains: A global approach to an integrated kinetic chain. *Medical Hypotheses, 110*, 90–96.

Doyon, J. K., Hajnal, A., Surber, T., Clark, J. D., & Kelty-Stephen, D. G. (2019). Multifractality of posture modulates multisensory perception of stand-on-ability. *PLoS One, 14*, e0212220.

Drenckhan, W., Hutzler, S., & Weaire, D. (2005). Foam physics: The simplest example of soft condensed matter. In: *AIP Conference Proceedings* (pp. 22–28).

Dupre, J., & Nicholson, D. J. (2018). A manifesto for a processual philosophy of biology. In D. Nicholson & J. Dupre (Eds.), *Everything flows: Towards a processual philosophy of biology* (pp. 3–45). Oxford University Press.

Eager, D., Pendrill, A. M., & Reistad, N. (2016). Beyond velocity and acceleration: Jerk, snap and higher derivatives. *European Journal of Physics, 37*, 065008.

Edmondson, A. C. (2007). *A Fuller explanation: The synergetic geometry of R. Buckminster Fuller*. Emergent World Press.

Edwards, S. A., Wagner, J., & Gräter, F. (2012). Dynamic prestress in a globular protein. *PLoS Computational Biology, 8*, 1002509.

Eichenauer, M. F., & Lordick, D. (2019). How Platonic and Archimedean solids define natural equilibria of forces for tensegrity. *FME Transactions, 47*, 234–244.

Ellers, O., Ellers, K.-I., Johnson, A. S., Po, T., Heydari, S., Kanso, E., & McHenry, M. J. (2024). Soft skeletons transmit force with variable gearing. *Journal of Experimental Biology, 227*, 246901.

Farina, S. C., Kane, E. A., & Hernandez, L. P. (2019). Multifunctional structures and multistructural functions: Integration in the evolution of biomechanical systems. *Integrative and Comparative Biology, 59*, 338–345.
Fraldi, M., Palumbo, S., Carotenuto, A. R., Cutolo, A., Deseri, L., & Pugno, N. (2019). Buckling soft tensegrities: Fickle elasticity and configurational switching in living cells. *Journal of the Mechanics and Physics of Solids, 124*, 299–324.
Fraternali, F., Stehling, N., Amendola, A., Anrango, B. A. T., Holland, C., & Rodenburg, C. (2020). Tensegrity modelling and the high toughness of spider dragline silk. *Nanomaterials, 10*, 1–15.
Freutel, M., Schmidt, H., Dürselen, L., Ignatius, A., & Galbusera, F. (2014). Finite element modeling of soft tissues: Material models, tissue interaction and challenges. *Clinical Biomechanics, 29*, 363–372.
Fuller, R. B. (1982). *Synergetics: Explorations in the geometry of thinking*. Collier Macmillan.
Fullerton, B. D. (2018). Biotensegrity: How ultrasound diagnostics guide regenerative orthopedic therapies to restore biomechanical function. In I. Kohlstadt & K. Cintron (Eds.), *Metabolic therapies in orthopedics* (2nd ed., pp. 91–109). Taylor and Francis.
Fultot, M., Frazier, P. A., Turvey, M. T., & Carello, C. (2019). What are nervous systems for? *Ecological Psychology, 31*, 218–234.
Gatt, R., Wood, M., Gatt, A., Zarb, F., Formosa, C., Azzopardi, K., Casha, A., Agius, T., Schembri-Wismayer, P., Attard, L., & Al, E. (2015). Negative Poisson's ratios in tendons: An unexpected mechanical response. *Acta Biomaterialia, 24*, 201–208.
Glazier, P. S., & Davids, K. (2009). The problem of measurement indeterminacy in complex neurobiological movement systems. *Journal of Biomechanics, 42*, 2694–2696.
Gómez-Jáuregui, V., Carillo-Rodriguez, A., Manchado, C., & Lastra-Gonzalez, P. (2023). Tensegrity applications to architecture, engineering and robotics: A review. *Applied Sciences, 13*, 8669.
Graner, F., & Riveline, D. (2017). 'The forms of tissues, or cell-aggregates': van der Gucht, J. (2018). Grand challenges in soft matter physics. *Frontiers in Physics, 6*, 87.
Guimberteau, J.-C., Sawaya, E. T., & Armstrong, C. (2024). Intra-tissue endoscopic exploration under the skin of living patients: An invitation to re-think the nature of human living matter. *Journal of Bone Research and Orthopaedic Surgery, 2*, 103.
Hakkak, F., Jabalameli, M., Rostami, M., & Parnianpour, M. (2015). The tibiofemoral joint gaps—An arthroscopic study. *SDRP Journal of Biomedical Engineering, 1*, 1–6.
Harris, A. K. (2018). The need for a concept of shape homeostasis. *Bio Systems, 173*, 65–72.
Hayati, H., Eager, D., Pendrill, A.-M., & Alberg, H. (2020). Jerk within the context of science and engineering—A systematic review. *Vibration, 3*, 371–409.
Heartney, E. (2009). *Kenneth Snelson: Forces made visible*. Hard Press Editions.
Hirst, L. (2013). *Fundamentals of soft matter science*. CRC Press.
Humphrey, J. D. (2003). Continuum biomechanics of soft biological tissues. In *Proceedings of the Royal Society A: Mathematical, Physical and Engineering Sciences* (Vol. 459, pp. 3–46).
Ingber, D. E. (2023). From tensegrity to human organs-on-chips: Implications for mechanobiology and mechanotherapeutics. *Biochemical Journal, 480*, 243–257.
Ingber, D. E., Madri, J., & Jamieson, J. (1981). Role of basal lamina in neoplastic disorganization of tissue architecture. *Proceedings of the National Academy of Sciences, 78*, 3901–3905.
Ingber, D. E., Wang, N., & Stamenović, D. (2014). Tensegrity, cellular biophysics, and the mechanics of living systems. *Reports on Progress in Physics. Physical Society, 77*, 046603.
Jung, E., Ly, V., Cheney, C., Cessna, N., Ngo, M. L., & Castro, D. (2021). Design, construction and validation of a proof of concept flexible-rigid mechanism emulating human leg behavior. *Applied Sciences, 11*, 9351.
Juran, C. M., Dolwick, M. F., & McFetridge, P. S. (2013). Shear mechanics of the TMJ disc: Relationship to common clinical observations. *Journal of Dental Research, 92*, 193–198.
Kiely, J., & Collins, D. J. (2016). Uniqueness of human running coordination: The integration of modern and ancient evolutionary innovations. *Frontiers in Psychology, 7*, 262.
Kim, W., & Turvey, M. T. (2018). The knee tensegrity system manages the balance between tension and compression. *Journal of Yoga and Physiotherapy, 6*, 56–59.

Kim, W., Thambyah, A., & Broom, N. (2012). Does prior sustained compression make cartilage-on-bone more vulnerable to trauma? *Clinical Biomechanics, 27*, 637–645.

Kim, P.-T., Isogai, S., Murakami, G., Wada, T., Aoki, M., Yamashita, T., & Ishii, S. (2002). The lateral collateral ligament complex and related muscles act as a dynamic stabilizer as well as a static supporting structure at the elbow joint: An anatomical and experimental study. *Okajimas Foka Anatomy Japan, 79*, 55–62.

Kirkby, R. (1975). The probable reality behind structural integration. *Journal of Structural Integration.*

Krause, J., & Lichtenstein, C. (eds.). (1999). *Your private sky—R. Buckminster Fuller: the art of design science.* Lars Müller.

Kroto, H. (1990). C60, fullerenes, giant fullerenes and soot. *Pure and Applied Chemistry, 62*, 407–415.

Lamarck, J. B. (1963). *Zoological philosophy: An exposition with regard to the natural history of animals.* Hafner.

Levin, S. M. (1981). The icosahedron as a biologic support system. In *34th Annual Conference Alliance for Engineering in Medicine & Biology* (p. 404).

Levin, S. M. (1982). Continuous tension, discontinuous compression. *Bulletin of Structural Integration, 8.*

Levin, S. M. (1997). Putting the shoulder to the wheel: A new biomechanical model for the shoulder girdle. *Journal of Biomedical Sciences Instrumentation, 33*, 412–417.

Levin, S. M. (2002). The tensegrity-truss as a model for spine mechanics: Biotensegrity. *Journal of Mechanics in Medicine and Biology, 2*, 375–388.

Levin, S. M. (2007). A suspensory system for the sacrum in pelvic mechanics: Biotensegrity. In A. Vleeming, V. Mooney, & R. Stoedkart (Eds.), *Movement, stability and lumbopelvic pain* (2nd ed.). Elsevier.

Levin, S. M. (2015). Tensegrity: The new biomechanics. In M. Hutson & A. Ward (Eds.), *Oxford textbook of Musculoskeletal medicine* (2nd ed., pp. 150–162). Oxford University Press.

Levin, S. M., & Madden, M. A. (2005). *In vivo observation of articular surface contact in knee joints.* Unpublished. www.biotensegrity.com

Levin, S. M., & Scarr, G. (2022). Biotensegrity and the mechanics of fascia. In R. Schleip, P.A. Huijing, C. Stecco & M. Driscoll (Eds.), *Fascia: The tensional network of the human body* (2nd ed.). Churchill Livingstone Elsevier.

Levin, S. M., & Lowell de Solórzano, S. (2024). Bouncing bones—Ancient wisdom meets modern science in a new take on locomotion. *Frontiers in Physiology, 15*, 1432410.

Levin, S. M. L., de Solórzano, S., & Scarr, G. (2017). The significance of closed kinematic chains to biological movement and dynamic stability. *Journal of Bodywork and Movement Therapies, 21*, 664–672.

Lin, D. C., McGowan, C. P., Blum, K. P., & Ting, L. H. (2019). Yank: The time derivative of force is an important biomechanical variable in sensorimotor systems. *Journal of Experimental Biology, 222*, 181414.

Lord, E. A. (2002). Helical structures: The geometry of protein helices and nanotubes. *Structural Chemistry, 13*, 305–314.

Lord, E. A., & Ranganathan, S. (2001). Sphere packing, helices and the polytope {3, 3, 5}. *European Physics Journal D, 15*, 335–343.

Maina, J. N. (2017). Pivotal debates and controversies on the structure and function of the avian respiratory system: Setting the record straight. *Biological Reviews, 92*, 1475–1504.

Malda, J., de Grauw, J. C., Benders, K. E. M., Kik, M. J. L., van de Lest, C. H. A., Creemers, L. B., Dhert, W. J. A., & van Weeren, P. R. (2013). Of mice, men and elephants: The relation between articular cartilage thickness and body mass. *PLoS One, 8*, e57683.

Mangalam, M., Chen, R., McHugh, T. R., Singh, T., & Kelty-Stephen, D. G. (2020). Bodywide fluctuations support manual exploration: Fractal fluctuations in posture predict perception of heaviness and length via effortful touch by the hand. *Human Movement Science, 69*, 102543.

Mao, Y., & Wickstrom, S. A. (2024). Mechanical state transitions in the regulation of tissue form and function. *Nature Reviews Molecular Cell Biology., 25*, 654–670.
Masi, A. T., & Hannon, J. C. (2009). Human resting muscle tone (HRMT): Narrative introduction and modern concepts. *Journal of Bodywork and Movement Therapies, 12*, 320–332.
Melnyk, A., & Pitti, A. (2018). Synergistic control of a multi-segments vertebral column robot based on tensegrity for postural balance. *Advanced Robotics, 32*, 850–864.
Mirabet, V., Das, P., Boudaoud, A., & Hamant, O. (2011). The role of mechanical forces in plant morphogenesis. *Annual Review of Plant Biology, 62*, 365–385.
Muller, M. (1996). A novel classification of planar four-bar linkages and its application to the mechanical analysis of animal systems. *Philosophical transactions of the Royal Society of London. Series B, Biological sciences, 351*, 689–720.
Muñoz, M. M. (2019). The evolutionary dynamics of mechanically complex systems. *Integrative and Comparative Biology, 59*, 705–715.
Nicholson, D. J. (2018). Reconceptualizing the organism: From complex machine to flowing stream. In D. Nicholson & J. Dupré (Eds.), *Everything moves: Towards a processual philosophy of biology* (pp. 139–166). Oxford University Press.
Nicholson, D. J. (2019). Is the cell really a machine? *Journal of Theoretical Biology, 477*, 108–126.
Oh, C. L., Keong, C. K., Nishimura, T., & Jae-yeol, K. (2022). Multi-directional shape change analysis of biotensegrity model mimicking human spine curvature. *Applied Sciences, 12*, 2377.
Olsen, A. M. (2019). A mobility-based classification of closed kinematic chains in biomechanics and implications for motor control. *Journal of Experimental Biology, 222*, 195735.
Pardehshenas, H., Maroufi, N., Sanjari, M. A., Parnianpour, M., & Levin, S. M. (2014). Lumbopelvic muscle activation patterns in three stances under graded loading conditions: Proposing a tensegrity model for load transfer through the sacroiliac joints. *Journal of Bodywork and Movement Therapies, 18*, 633–642.
Parker, K. K., & Ingber, D. E. (2007). Extracellular matrix, mechanotransduction and structural hierarchies in heart tissue engineering. *Philosophical Transactions of the Royal Society B: Biological Sciences, 362*, 1267–1279.
Pascal, R., & Pross, A. (2015). Stability and its manifestation in the chemical and biological worlds. *Chemical Communications, 51*, 16160–16165.
Persaud-Sharma, D., & O'Leary, J. P. (2015). Fibonacci series, Golden proportions, and the human biology. *Austen Journal of Surgery, 2*, 1–6.
Piolanti, N., Polloni, S., Bonicoli, E., Giuntoli, M., Scaglione, M., & Indelli, P. F. (2018). Giovanni Alfonso Borelli: The precursor of medial pivot concept in knee biomechanics. *Joints, 6*, 167–172.
Primus, M. A. (2022). Social tensegrity and family systems. *Journal of Human Behavior in the Social Environment, 34*, 210–220.
Profeta, V. L. S., Turvey, M. T., & Carello, C. (2020). Goal-directed action and the architecture of movement organization. In M. L. Latash (Ed.), *Berstein's construction of movements*. Routledge.
Reilly, C. B., & Ingber, D. E. (2018). Multi-scale modeling reveals use of hierarchical tensegrity principles at the molecular, multi-molecular, and cellular levels. *Extreme Mechanics Letters, 20*, 21–28.
Reznikov, N., Chase, H., Ben Zvi, Y., Tarle, V., Singer, M., Brumfeld, V., Shahar, R., & Weiner, S. (2016). Inter-trabecular angle: A parameter of trabecular bone architecture in the human proximal femur that reveals underlying topological motifs. *Acta Biomaterialia, 44*, 65–72.
Rimoli, J. J., & Pal, R. K. (2017). Mechanical response of 3-dimensional tensegrity lattices. *Composites Part B: Engineering, 115*, 30–42.
Rivier, N., Sadoc, J. F., & Charvolin, J. (2016). Phyllotaxis: A framework for foam topological evolution. *European Physical Journal E, 39*, 1–11.
Robbie, D. L. (1977). Tensional forces in the human body. *Orthopaedic Review, 6*, 45–48.
Russell, B. (2009). *The scientific outlook*. Routledge.
Sadoc, J. F., & Rivier, N. (2000). Boerdijk-Coxeter helix and biological helices as quasicrystals. *Materials Science and Engineering: A, 294–296*, 397–400.

Saxena, A., Lipson, H., & Valero-Cuevas, F. J. (2012). Functional inference of complex anatomical tendinous networks at a macroscopic scale via sparse experimentation. *PLoS Computational Biology, 8*, e1002751.
Scarr, G. (2008). A model of the cranial vault as a tensegrity structure and its significance to normal and abnormal cranial development. *International Journal of Osteopathic Medicine, 11*, 80–89.
Scarr, G. (2012). A consideration of the elbow as a tensegrity structure. *International Journal of Osteopathic Medicine, 15*, 53–65.
Scarr, G. (2016). Fascial hierarchies and the relevance of crossed-helical arrangements of collagen to changes in the shape of muscles. *Journal of Bodywork and Movement Therapies, 20*, 377–387.
Scarr, G. (2018). *Biotensegrity: The structural basis of life* (2nd ed.). Handspring.
Scarr, G. (2020). Biotensegrity: What is the big deal? *Journal of Bodywork and Movement Therapies, 24*, 134–137.
Scarr, G. (2025). Movement is a whole-body issue. In E. Larkam, & M, Black (Eds.), *Pilates applications for health conditions* (Vol. 1). Handspring.
Scarr, G., & Harrison, H. (2016). Resolving the problems and controversies surrounding temporo-mandibular mechanics. *Journal of Applied Biomedicine, 14*, 177–185.
Scarr, G., & Harrison, H. (2017). Examining the temporo-mandibular joint from a biotensegrity perspective: A change in thinking. *Journal of Applied Biomedicine, 15*, 55–62.
Scarr, G., Blyum, L., Levin, S. M., & de Solórzano, S. L. (2024). Moving beyond Vesalius: Why anatomy needs a mapping update. *Medical Hypotheses, 183*, 111257.
Schrödinger, E. (1944). *What is life?* Cambridge University Press.
Serón, F. J., Zaldívar, Á., Blesa, A., Celani, G., & Magallón, J. (2022). A theoretical reflection on smart shape modeling. *International Journal on Interactive Design and Manufacturing, 16*, 821–839.
Sitharam, M., & Agbandje-Mckenna, M. (2006). Modeling virus self-assembly pathways: Avoiding dynamics using geometric constraint decomposition. *Journal of Computational Biology, 13*, 1232–1265.
Skelton, R. E., Helton, W. W., Chan, W., Pinaud, J.-P., & Adhikari, R. (2001). An introduction to tensegrity structures. In *Proceedings of the 40th IEEE Conference on Decision and Control* (pp. 1–6).
Snelson, K. (1996). Snelson on the tensegrity invention. *International Journal of Space Structures, 11*, 43–48.
Stamenović, D., & Smith, M. L. (2020). Tensional homeostasis at different length scales. *Soft Matter, 16*, 6946–6963.
Standring, S. (2005). *Gray's anatomy* (39th ed.). Elsevier Churchill Livinstone.
Tadeo, I., Berbegall, A. P., Escudero, L. M., Álvaro, T., & Noguera, R. (2014). Biotensegrity of the extracellular matrix: Physiology, dynamic mechanical balance, and implications in oncology and mechanotherapy. *Frontiers in Oncology, 4*, 39.
Terayama, K., Takei, T., & Nakada, K. (1980). Joint space of the human knee and hip joint under a static load. *Engineering in Medicine, 9*, 67–74.
Thompson, D. W. (1917). *On growth and form*. Cambridge University Press.
Turvey, M. T., & Fonseca, S. T. (2014). The medium of haptic perception: A tensegrity hypothesis. *Journal of Motor Behavior, 46*, 143–187.
Valero-Cuevas, F. J., Yi, J.-W., Brown, D., McNamara, R. V., Paul, C., & Lipson, H. (2007). The tendon network of the fingers performs anatomical computation at a macroscopic scale. *IEEE Transactions on Biomedical Engineering, 54*, 1161–1166.
Van Essen, D. C. (2023). Biomechanical models and mechanisms of cellular morphogenesis and cerebral cortical expansion and folding. *Seminars in Cell and Developmental Biology, 140*, 90–104.
Vesalius, A. (1543). *De humani corporis fabrica*. University of Padua.
Wilke, J., Schleip, R., Yucesoy, C. A., & Banzer, W. (2018). Not merely a protective packing organ? A review of fascia and its force transmission capacity. *Journal of Applied Physiology, 124*, 234–244.

Wilson, J., & Kiely, J. (2016). The multi-functional foot in athletic movement: Extraordinary feats by our extraordinary feet. *Human Movement, 17*, 15–20.

Witztum, A., & Wayne, R. (2014). Fibre cables in the lacunae of Typha leaves contribute to a tensegrity structure. *Annals of Botany, 113*, 789–797.

Yang, Q., Cao, M., & Anderson, B. D. O. (2019). Growing super stable tensegrity frameworks. *IEEE Transactions on Cybernetics, 49*, 2524–2535.

Ziauddin, M., & Ali, B. (2013). An ultrasonic study on elasticity of animal soft tissues. *International Journal of Science, Environment and Technology, 2*, 1257–1261.

Chapter 8
Shape Control for Biotensegrities

Chai Lian Oh, Kok Keong Choong, and Toku Nishimura

Abstract Biotensegrity, which integrates tensegrity principles with biological structures, describes how living organisms efficiently distribute mechanical forces to accommodate stresses. This chapter provides an overview of biotensegrity in anatomy and physiology, spanning from the macro to the micro level of a living organism. The chapter details the development of a biotensegrity model that replicates the tapered vertebral bodies and natural curvature of the human spine. A form-finding method for generating an n-stage three-strut spine model using static equilibrium equations is presented. Shape change strategies are explored using Sequential Quadratic Programming (SQP), enabling identified monitored nodes to move from their initial positions to target coordinates by adjusting cable lengths. Additionally, an obstacle avoidance strategy based on the Potential Method facilitates flexible motion, allowing the model to reach target positions while navigating around obstacles. Comparative results highlight the influence of obstacle avoidance considerations. The chapter concludes with insights into future research directions, emphasizing biotensegrity's potential applications in biomechanics, robotics, and structural design.

8.1 Overview of Biotensegrity

Biotensegrity principle, developed by Levin in the 1980s applies the principle of human-made tensegrity structures to the field of biology. Biotensegrity describes how living organisms accommodate stresses effectively through the distribution of

C. L. Oh (✉)
School of Civil Engineering, College of Engineering, Universiti Teknologi MARA, 40450 Shah Alam, Selangor, Malaysia
e-mail: chailian@uitm.edu.my

K. K. Choong
School of Civil Engineering, Engineering Campus, Universiti Sains Malaysia, 14300 Nibong Tebal, Pulau Penang, Malaysia

T. Nishimura
Department of Architecture, College of Architecture, Kanazawa Institute of Technology, Hakusan 924-0838, Japan

F. Fraternali and J. J. Rimoli (eds.), *Tensegrity Systems*, CISM International Centre for Mechanical Sciences 2, https://doi.org/10.1007/978-3-031-82283-4_8

mechanical forces over specific pathways. The external impacts generate responses locally and at a distance that eventually contribute to the global deformation of the biological system. These force transmission pathways comprise isolated compressive struts and continuous tensional muscles, making the living organisms work in contraction and expansion. These multiple structural elements are balanced and integrated within the organism to form a stable self-equilibrated organisation that may spontaneously reconfigure through elementary self-assemblage processes once disturbed. The hierarchical biotensegrities perform multiple dynamic functions simultaneously, such as sense, signal, and feedback, consequently changing shapes and mechanics at different scales, similar to those in a biological system.

Section 8.1 presents an overview of biotensegrity in anatomy and physiology, followed by a description of a biotensegrity model that mimics the human spine in Sect. 8.2. Sections 8.3 and 8.4 detail the form-finding method used to generate an n-stage three-strut spine biotensegrity model. Sections 8.5 and 8.6 discuss shape change strategies using static equilibrium equations, comparing results with and without considerations for obstacle avoidance. Finally, conclusions and future recommendations are provided.

Employed Notation. This chapter uses the following symbols for key variables: nodal coordinates vector $\mathbf{x}$; nodal forces vector $\mathbf{F}$; axial forces vector $\mathbf{n}$; equilibrium matrix $\mathbf{B}$; geometrical stiffness matrix ${}^t\mathbf{k}_g$; linear stiffness matrix ${}^t\mathbf{k}_e$, tangent stiffness matrix ${}^t\boldsymbol{K}$; incremental axial force vector ${}^t\dot{\mathbf{n}}$; incremental nodal coordinates vector ${}^t\dot{\mathbf{x}}_G$; and force elongation vector ${}^t\dot{\mathbf{l}}$.

Anatomy. Human organ systems are structured in accordance with biotensegrity architecture, especially the skeletal system, such as the spine, shoulder, elbow, pelvis, distal radio-ulnar joint, and cranial vault (Hagert & Hagert, 2010; Levin, 2002, 2007; Scarr, 2008, 2012). Human organ systems are possibly assembled from sets of compressive and tensile components. The rigid components, like load-bearing bones, are generally described as compressive components that are held in position with soft tensile components, such as muscles, tendons, ligaments, and fascia net. This would explain how the skeletal systems are suspended in the body, simultaneously accommodating their large mass without collapse (i.e. elephant, whale) and maintaining their mobility just by the action of muscles. Furthermore, several examples of bones not in contact in human anatomy are the separation of vertebrae by intervertebral discs, the existence of a gap between the humerus, ulna, and radius at the synovial joint in the elbow, and the isolation of curved cranial bones by sutures. Therefore, the integration of discontinuous compressive components and continuous tension components may be one of the key features in human anatomy, as described through the biotensegrity principle.

In addition to the coordination between a few solid materials, such as the discontinuous struts within fibres, tensegrity is exemplified by an inflated balloon, camp tent, bicycle wheel, and spider web (Levin, 2002). Such structure is maintained under prestressed conditions, with the tensile stresses in the system before any application of external loadings. At the organ level, Levin (2002) described organs such as the heart, lungs, and urinary bladder as prestressed structures. They contract and expand like a "tensegrity pump". Under contraction, the inner pressure increases, and the shape rigidity is enhanced. Maina (2007) observed that the lung with tightly packed parabronchi (compression) utilises a hexagonal geodesic design to enhance the avian lung's rigidity. The intertwined atrial muscles, the web-like laid-out interparabronchial septa, and tightly packaged air- and blood capillaries (tension) efficiently transmit and dissipate stresses within the parabronchi.

The body of research explains the significant application of biotensegrity in anatomy, even at the microscale and nanoscale levels. The existence of tension (prestress) in living organisms has been identified at the tissue level (Ingber, 2003a), in the cerebral cortex (Essen, 1997), and at the cellular level, such as the cytoskeleton (Ingber, 2003b). Dr. Ingber conducted numerous works on the principle of biotensegrity in mechanobiology, especially the cellular tensegrity model (Ingber, 2023). Cytoskeleton maintains the shape of cells by utilising internal filamentous biopolymers such as actin microfilaments, intermediate filaments (i.e. tensional elements), and microtubules (i.e. compressive elements). The external sources, especially the extracellular matrix (ECM, the stiffer complexes surrounding the cells), also act as a compressive element in providing structural support to the cell (Ingber, 2003a). In addition, another smaller scale of the biotensegrity structure has been speculated to exist within the cytoskeleton. This indicates that the living organism is indeed a huge stable network consisting of multi-modular biotensegrities that hierarchically communicate and perform similar or multiple tasks towards a common goal.

Physiology. Global reorientation was observed in the mechanically stressed globular protein molecules and DNA structure (Edwards et al., 2012). At a higher hierarchical level, the disturbance of muscles easily transforms human skeletal posture to produce work and maintain mechanical stability (Levin, 2002). Ingber (2003a) has concluded that the rapid transmission of forces within the hierarchical system from the macro to the micro level of a living organism could further reinforce the entire network without disturbing any structural connectivity. This may be due to the selective pathways followed by the stiffer routes, which prevent the damage of any particular unit.

Another novel biotensegrity model in physiology is the sensing and feedback mechanism. The transportation of information through possible signalling pathways in multi-hierarchical elements in the biotensegrity model is well-known. This unique model successfully explains cellular mechanotransduction, a process for cells to convert external physical forces into the adaptation of intracellular biochemistry and gene activities (Ingber, 2010). For instance, once mechanoreceptors, particularly integrins, detect the application of mechanical forces at the cell surface, the focal adhesions instantly change in size, composition, and location. Ingber has also found that this signal transmission is further across the cytoskeletal filaments and concentrates deep into the nucleus. It consequently promotes the alteration of biochemistry inside and outside the cells, almost at the same time as the impact forces. Rapid and distant wave signal transmission through the connective intracellular elements is crucial for cell behaviour.

8.2 Biotensegrity Model Mimicking Human Spine

Despite being spongy, the tapered spinal vertebrae body efficiently maintains body weight and withstands external impacts. In addition to the unique architecture with sagittal curvature of the spinal column, the spine can avoid distress (i.e., exhaustion) by re-establishing new equilibrium positions. The dexterity in avoidance of distress in the system is mainly due to the harmonisation of the passive (i.e., spinal column), active (i.e., surrounding muscles and tendons) and control (i.e., neural centres) subsystems. It is noted that the spine always maintains the stability of the system even after disturbance by a self-structural reorganisation through many postures (Levin, 2002). Based on the above descriptions, a spine has a distinctive characteristic of flexibility and versatility in accommodating different movements. The principle of biotensegrity is suitable for explaining the mechanism of the compressive passive subsystem (i.e. spinal column) and tensional active subsystem (i.e., surrounding muscles and tendons) in the spine.

Vertebrae Anatomical Parameters. Figure 8.1a shows a typical human spine in sagittal and posterior views, whereas Fig. 8.1b shows typical vertebrae in superior and lateral views. Sagittal and posterior views are medical terms that refer to the appearance of the spine when viewed from the side and behind, respectively. A human spine consists of a total of twenty-two vertebrae from the cervical (vertebrae C1–C7), thoracic (vertebrae T1–T12) and lumbar regions (vertebrae L1–L5). Anatomical parameters were obtained from (Busscher et al., 2010), such as the vertebral end-plate width (EPW), vertebral central body height (VBHC), and intervertebral disc height (IDH) in the generation of the spine biotensegrity model.

Natural Sagittal Curvature. Figure 8.2 shows the established natural sagittal line for the spine biotensegrity models, considering the natural curvature, C7 plumb line, and locations of apex. The human spine has a natural curvature in sagittal view with (i) lordosis (concave curve) at cervical and lumbar regions and (ii) kyphosis (convex curve) at thoracic and sacral regions. Cervical lordosis of 20.1°, thoracic kyphosis of 34.5° and lumbar lordosis of 29.2° were measured as the angles between the upper end-plate of C3 and the lower end-plate of C7, the upper end-plate of T1 and the lower end-plate of T12 and the upper end-plate of L1 and the lower end-plate of L5, respectively (Busscher et al., 2010). The apex of the curvature is defined as the farthest point of a vertebra from the C7 plump line. The C7 plumb line is established by extending a vertical line from the centre of the C7 vertebral body to the posterior corner (at the back) of sacrum S1. The normal weight of the human body generally acts through the C7 plumb line. It is noted that there is an apex each within cervical lordosis, thoracic kyphosis and lumbar lordosis in a human spine. The apex of lumbar lordosis, AP1, is assumed at the intervertebral disc between L3-L4, whereas the apex of thoracic kyphosis, AP2, is assumed at vertebrae T7, the apex of cervical lordosis, AP3, is assumed at C5. Cobb method has been adopted to establish the curvature of the spine biotensegrity model (Vrtovec et al., 2009).

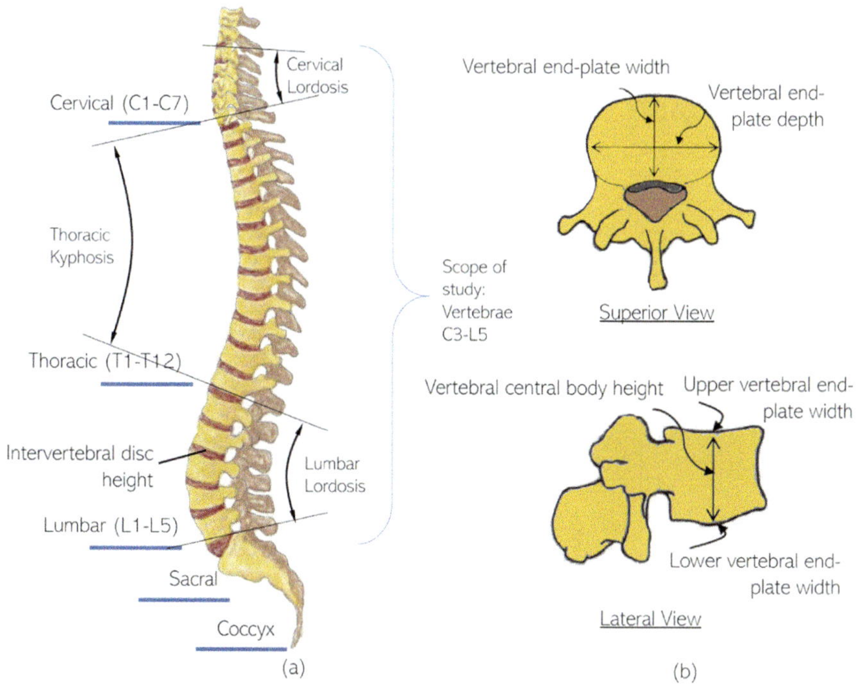

Fig. 8.1 **a** Human spine anatomy (lateral view), and **b** typical vertebrae (superior and lateral views) (Oh et al., 2022)

8.3 Generation of *n*-stage Three-Strut Spine Biotensegrity Model

Definition of Basic Cell. Figure 8.3 illustrates a typical triangular cell of the spine biotensegrity model. The geometry of a triangular cell is presented in the global Cartesian coordinate system O_o-$X_oY_oZ_o$. The position vector of nodal coordinates is denoted as **X**. The origins of the local Cartesian system O_{i-1}-$X_{i-1}Y_{i-1}Z_{i-1}$ and O_i-$X_iY_iZ_i$ are positioned at the centroid of lower and upper triangular surfaces of i_{th} ($i = 1, 2, \ldots, n$) triangular cell, respectively. The nodes at lower L_{iB}, upper triangular surfaces L_{iT} and the centroids of the top and bottom triangles O_i and $O_{i\text{-}1}$ of a triangular cell are defined in Eq. (8.1), with φ is the rotation of the triangular surface h is the height of the cell.

$$L_{iB} = \left[X_{6i-5}\, X_{6i-4}\, X_{6i-3}\right]$$
$$L_{iT} = \left[X_{6i-2}\, X_{6i-1}\, X_{6i}\right]$$

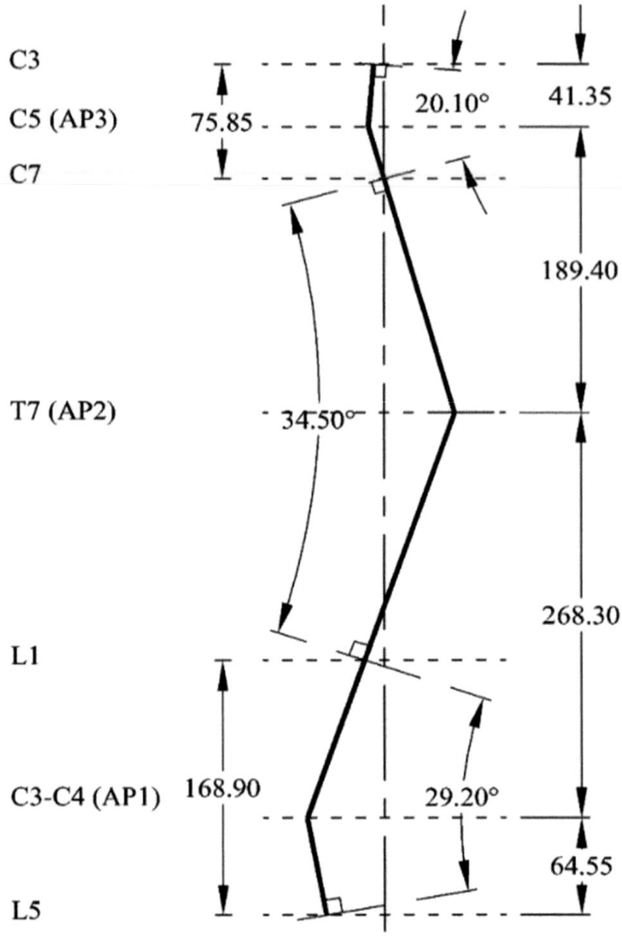

Fig. 8.2 Established curvature line mimicking the human spine (Oh et al., 2020)

$$O_i = \begin{Bmatrix} x_o \\ y_o \\ z_o \end{Bmatrix}$$

$$X_i = \begin{Bmatrix} x_i \\ y_i \\ z_i \end{Bmatrix} = \begin{Bmatrix} \left[(x_{in} - x_o)\cos\varphi - (y_{in} - y_o)\sin\varphi\right] + x_o \\ \left[(x_{in} - x_o)\sin\varphi + (y_{in} - y_o)\cos\varphi\right] + y_o \\ z_i \end{Bmatrix} \tag{8.1}$$

The nodal coordinates in plane XY of the spine biotensegrity models are determined based on triangle altitude, d and the vertex angle, θ of the lower and upper triangular surfaces in the local Cartesian system. The tapered section properties of a human spine are incorporated into the spine biotensegrity models (the width of the

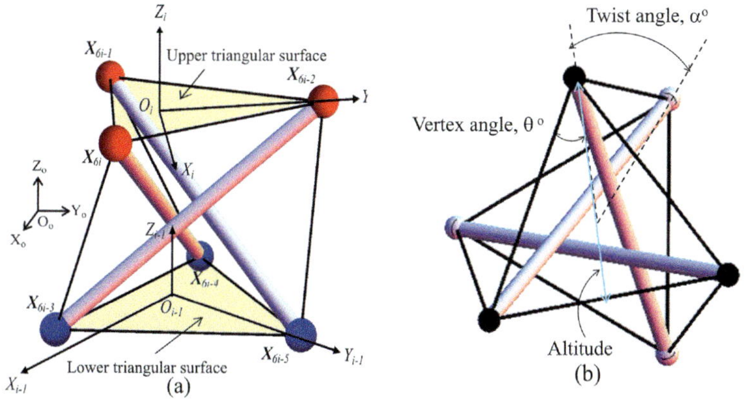

Fig. 8.3 Typical triangular cell: **a** perspective view, **b** plan view

vertebra gradually increases from cervical vertebrae to lumbar vertebrae). Reference is made to the EPW of vertebrae to characterise the widths of the spine biotensegrity models, and these widths are adopted as the triangle altitudes for the triangular surfaces in cells. While the accumulation of both VBHC and IDH represents the height of the spine biotensegrity model, the accumulation of IDH represents the saddle height between two independent cells.

Element Connectivity. Tensegrity is classified into class one, two and k depending on the number of struts at a node. Class one tensegrity mast was chosen for the spine biotensegrity model to mimic the stacking up of discontinuous compressive vertebral bodies of a human spine (Oh et al., 2020). The spine biotensegrity model consists of struts (i.e., vertebrae) that are connected solely by cables (i.e., muscles, intervertebral disc) and offers no contact in-between the struts (i.e. vertebrae are cushioned by intervertebral disc and surrounded by muscles).

Figure 8.4 shows the developed element connectivity chart for N-stage class-one spine biotensegrity models, inspired by Pugh's diamond pattern. The element connectivity at the second and third stages may be repeated to stack up to n triangular cells. The $(i+1)$th triangular cell is stacked up on the current ith triangular cell. Black bold lines represent the struts; red irregular dash-dotted lines represent saddle cables (connecting the interface between two triangular cells); the black regular dashed lines represent horizontal cables (joining the nodes only at the base and the top of the model). Zig-zag lines represent reinforcing cables at the first stage. Diagonal cables 1 (cyan colour) connect nodes L_{iB} and L_{i+1B}, diagonal cables 2 (pink colour) connect nodes L_{iT} and L_{i+1T}, and diagonal cables 3 (blue colour) connect nodes L_{iB} and L_{iT}.

Assemblage of Spine Biotensegrity Model. Figure 8.5 illustrates the assemblage process of a spine biotensegrity model with four triangular cells. Firstly, the triangular surfaces are placed at their centroids, coinciding with the established curvature line (see Fig. 8.2), with one at the base, two at each apex (a total of three apexes),

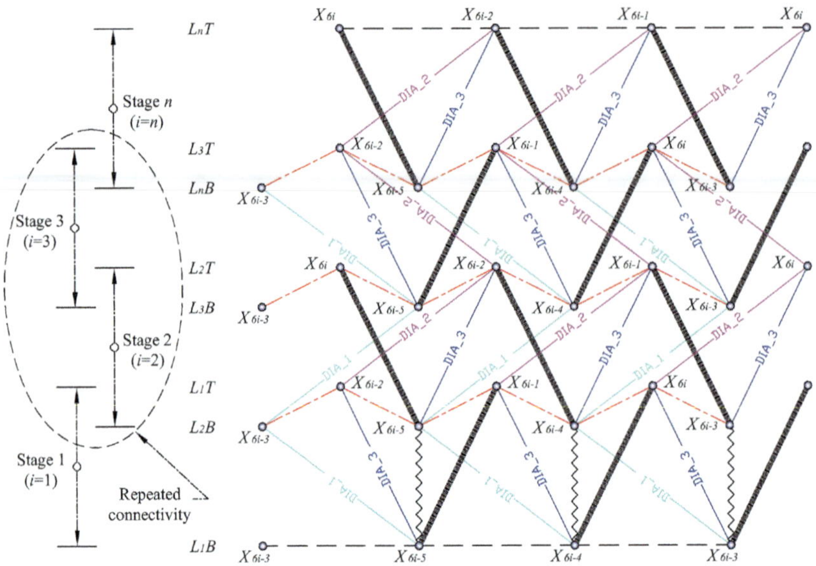

Fig. 8.4 Element connectivity of N-stage class one three-strut spine biotensegrity model (Oh et al., 2020)

and one at the top. The altitudes of the triangular surfaces based on the anatomical parameter EPW decrease along with the height, reflecting a human spine's taper form and slenderness. Note that nodes $X_{6i\text{-}5}$ and $X_{6i\text{-}2}$ of ith triangular cell are positioned on the global Y axis (say $x_{in} = 0$) prior to the twisting process (Fig. 8.5a). Next, the lower and upper triangular surfaces are rotated with angle φ at intra-module clockwise and counterclockwise (or otherwise) for odd- and even-number of ith triangular cell, respectively, resulting in a twist angle α (Fig. 8.5b). The rotation between the triangular surfaces forming the interface of two triangular cells at inter-module (say 60°) is the twist direction of the lower cell. Lastly, the nodes of the triangular surfaces are used to connect the struts and cables according to the developed element connectivity pattern in Fig. 8.4 (Fig. 8.5c). It is to be noted that the greater twisting angle may cause a collision between the elements, which need to be treated with care. The following section presents the formulation of the equations for searching for the self-equilibrated state of a spine biotensegrity model.

8.4 Form Finding of Self-Equilibrated Biotensegrity Model

The formulation of the self-equilibrated spine biotensegrity model is derived based on the basic formulations of pin-jointed structures with the following assumptions:

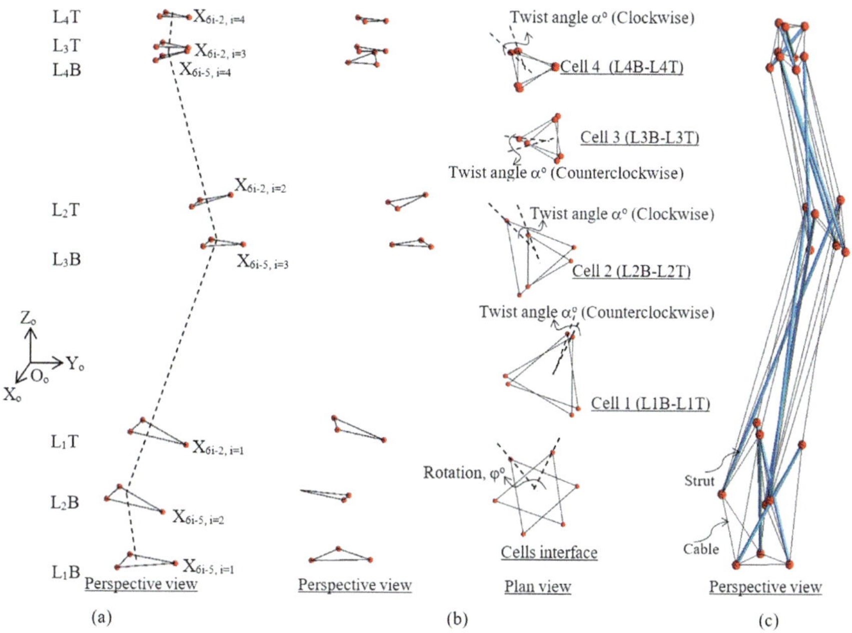

Fig. 8.5 Assemblage of a spine biotensegrity model **a** fitting of triangular surfaces perpendicular to the established curvature line, **b** twisting of cells in perspective and plan view, **c** positioning of struts and cables according to the element connectivity (Oh et al., 2020)

- Self-weight of struts and cables are neglected due to the reason that the stress resulted from self-weight is insignificantly small compared to the level of pre-stress (Ohsaki & Zhang, 2015).
- There is no external force acting on the models. It is necessary to ensure the self-equilibrium position, as shown under the topology of biotensegrity model is rational. Against any disturbance, the self-equilibrium and stability of the obtained topology should be retained.

The spine biotensegrity model consists of m elements, n_o nodes, n_c supports, and $n_u = 3(n_o - n_c)$ unconstrained degree of freedoms. Consider an element k with length L_T of the spine biotensegrity model connecting node i and j ($i < j$) and vectors of nodal coordinates of node i and j as $\mathbf{x}_i$ and $\mathbf{x}_j$, respectively. The vector of directional cosines of an element k is expressed as:

$$\boldsymbol{\lambda} = \frac{(\mathbf{x}_j - \mathbf{x}_i)}{L_T} \tag{8.2}$$

Considering static equilibriums at all nodes:

$$\mathbf{F} = \mathbf{B}\mathbf{n} \tag{8.3}$$

where $\mathbf{B}$ is a $n_u \times m$ matrix consisting of the vector of directional cosines of all elements in the spine biotensegrity model, $\mathbf{F}$ is a vector of nodal forces with size n_u and $\mathbf{n}$ is a vector of axial forces for all elements with size m. By making use of Moore–Penrose generalised inverse, the necessary and sufficient condition for the existence of a solution for Eq. (8.3) can be expressed:

$$\mathbf{F} = \mathbf{B}\mathbf{B}^{+}\mathbf{F} \tag{8.4}$$

which can be rewritten in the following form:

$$[\mathbf{I}_{n_u} - \mathbf{B}\mathbf{B}^{+}]\mathbf{F} = \mathbf{0} \tag{8.5}$$

where $\mathbf{I}$ is an identity matrix of order n_u and $\mathbf{B}^{+}$ is the Moore–Penrose generalised inverse of matrix $\mathbf{B}$. The singular value decomposition method is used to compute the Moore–Penrose generalised inverse of the matrix $\mathbf{B}$ with rank deficiency (i.e. rank $< n_u$). For the case of self-equilibrated state without external forces, $\mathbf{F} = 0$. Under such conditions, the solution for the vector of axial forces $\mathbf{n}$ is given as:

$$\mathbf{n} = \boldsymbol{B}^{+}\boldsymbol{B}]\boldsymbol{\beta} \tag{8.6}$$

where $\boldsymbol{\beta}$ is a vector of arbitrary coefficient of size m. Denoting matrix $[\mathbf{I}_m - \mathbf{B}^{+}\mathbf{B}]$ in Eq. (8.5) as matrix $\mathbf{G}$ of size $m \times m$ and applying the eigenvector basis decomposition method, the matrix $\mathbf{G}$ can be expressed:

$$\mathbf{G} = \boldsymbol{\Phi}\boldsymbol{\Lambda}\boldsymbol{\Phi}^{T} \tag{8.7}$$

where $\boldsymbol{\Lambda}$ is a diagonal matrix whose diagonal elements Λ_i $(i = 1, 2, \ldots, m)$ correspond to eigenvalues of matrix $\mathbf{G}$; $\boldsymbol{\Phi}$ is an orthogonal matrix whose ith column $\boldsymbol{\Phi}_i$ (i = 1, 2, …, m) corresponds to eigenvector associated with eigenvalue Λ_i. Since matrix $\mathbf{G}$ is an orthogonal projection matrix, the eigenvalues Λ_i return the value of either 0 or 1. The eigenvectors $\boldsymbol{\Phi}_i$ correspond to vectors of independent self-equilibrium stress modes. It is noted that the number of vectors of independent self-equilibrium stress modes is equal to the number of non-zero eigenvalues of matrix $\mathbf{G}$ or rank[$\mathbf{G}$]. The following expression for the vector of axial forces $\boldsymbol{n}$ can be obtained:

$$\boldsymbol{n} = \boldsymbol{\Phi}\boldsymbol{\Lambda}\boldsymbol{\beta}* \tag{8.8}$$

where $\boldsymbol{\beta}^* = \boldsymbol{\Phi}^{\mathrm{T}}\boldsymbol{\beta}$. Assuming that the spine biotensegrity model has q number of independent self-equilibrium stress modes, Eq. (8.8) can be further rewritten as:

$$\mathbf{n} = \sum_{i=1}^{q} \beta_i^* \boldsymbol{\Phi}_i \tag{8.9}$$

where β_i* is ith component of vector $\boldsymbol{\beta}$* which corresponds to the vector of coefficients for the linearly independent self-equilibrium stress modes.

The solution of the vector of axial force $\mathbf{n}$ is dependent on the combination of coefficients β_i* if the self-equilibrium stress modes are more than one, i.e. $q > 1$. In addition to checking against the yielding of cable elements, compressive forces of strut elements are checked against Euler's buckling load. The applied inequality constraints ensure that the axial forces for all the cable and strut elements are within their upper bound axial force limits and satisfy the elastic condition of the material properties. The set of β_i* ($i = 1, 2, 3,.., q$) determines the possible solution for $\mathbf{n}$, which satisfies the inequality constraints as follows:

$$\begin{aligned} 0 \le n_c \le \sigma_c A_c \\ -\frac{\pi^2 E_s I_s}{l_s^2} \le n_s \le 0 \end{aligned} \tag{8.10}$$

where n_c, E_c and A_c are axial forces, yield stress and cross-sectional area for cable elements, respectively; n_s, E_s, I_s, L_s, σ_s and A_s are axial forces, Young modulus, moment inertia of section, length, yield stress and cross-sectional area for strut. A numerical strategy for the determination of β_i* in order to obtain a possible self-equilibrated spine biotensegrity models is described in Oh et al., (2020).

Many form-finding methods for tensegrity structure have been proposed (Tibert and Pellegrino, 2011). A review of methods for the determination of feasible self-stress states for problems of tensegrity structures where $q > 1$ has been presented together with the relevant publications (Fraddosio et al., 2021). These methods have made use of optimisation techniques, mixed-integer programming strategies, spectral decomposition, stiffness matrix evaluation and numerical iterative procedures. A method which has made use of the concept of distributed statical indeterminacy (Zhou et al., 2015), with only a single singular value decomposition of equilibrium matrix $\mathbf{B}$ (see Eq. 8.3) has been proposed in Fraddosio et al., (2021) for the determination of a feasible self-stress state (vector $\mathbf{n}$ in Eq. 8.9). Input required for the proposed method are (a) connectivity relations between the elements and their types (cables and struts) and (b) nodal coordinates of the joints of the tensegrity structures. The outcome obtained through the proposed method is a linear combination of independent self-equilibrated modes (vector $\boldsymbol{\Phi}_\mathbf{i}$ in Eq. 8.9) consistent with the flexibility properties of the tensegrity structures. A step-by-step procedure has also been outlined whereby satisfaction of a unilateral condition, i.e. tension in cable and compression in strut, has been achieved through the specification of a test vector $\mathbf{n}_p$. However, the determination of self-equilibrium stresses in shape-changing tensegrity models has not been thoroughly investigated.

8.5 Shape Change Strategies of Biotensegrities: Static Equilibrium Equations

Terminology. The shape change involves four main states: initial, deformed, shape change and target states (Fig. 8.6). At the initial state, the tower is self-equilibrium and stands freely without constraints and external forces. The support condition is assigned at the deformed state, and the tower experiences linear elastic deformation under self-weight. The tower undergoes shape change through forced cable elongation at the shape change state. Finally, at the target state (or final state), the monitored nodes reach the prescribed target coordinates.

In a tensegrity model, consider an element k (where $k = 1, 2,.., m$; m = total number of elements) connects node i and j ($i < j$) in a Cartesian coordinate system. Figure 8.7 illustrates the configurations of element k of tensegrity model during shape change analysis. The equilibrium configurations of element k are evaluated at discrete steps: $t = 0$ (initial state), $t = 1$ (deformed state), step $t = \mathrm{i}$ (shape change state, with $i = 2, \ldots, N$), and $t = N$ (target state) where N is the total number of incremental steps. The nodal coordinates of nodes i and j at these steps are denoted as ${}^{0}x_i{}^{T}, {}^{1}x_i{}^{T}, {}^{i}x_i{}^{T}, {}^{N}x_i{}^{T}$ and ${}^{0}x_j{}^{T}, {}^{1}x_j{}^{T}, {}^{i}x_j{}^{T}, {}^{N}x_j{}^{T}$, respectively. Element k experiences elastic elongation (u), increment length due to elastic elongation $\mathring{u}$ and forced elongation $\dot{I}$ at specific steps. For a constrained (undeformed) cable, the forced elongation is zero as the cable length remains constant throughout the analysis. In the shape change analysis, certain nodes, called monitored nodes, are chosen from the unconstrained nodes to reach prescribed target coordinates $\overline{\mathbf{x}}$. Figure 8.8 explains the terminology of nodal coordinates of monitored nodes during the shape change analysis. The nodal coordinates of these monitored nodes at the previous and current

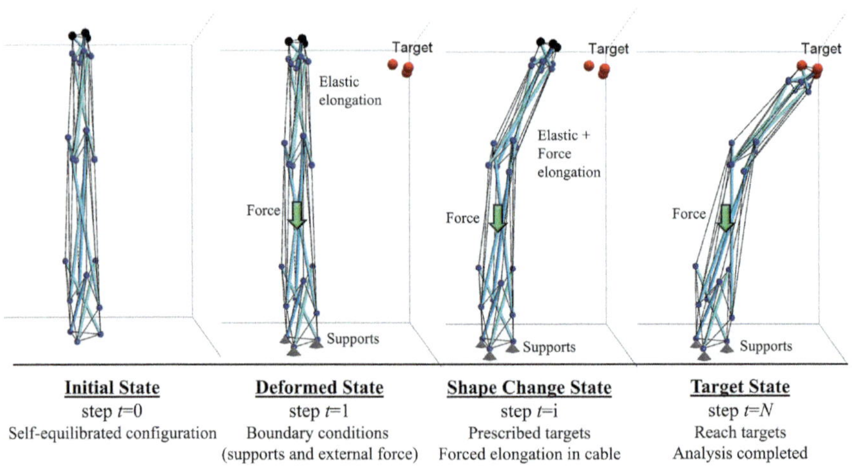

Fig. 8.6 Main states in shape change analysis

steps are denoted as ${}^{t-1}x$' and ${}^{t}x$', respectively, where *t*-1 and *t* represent the previous and current steps in the analysis.

Incremental Equilibrium Equations. The static incremental equilibrium equation for the entire structure could be obtained by differentiation of Eq. (8.2) as follows:

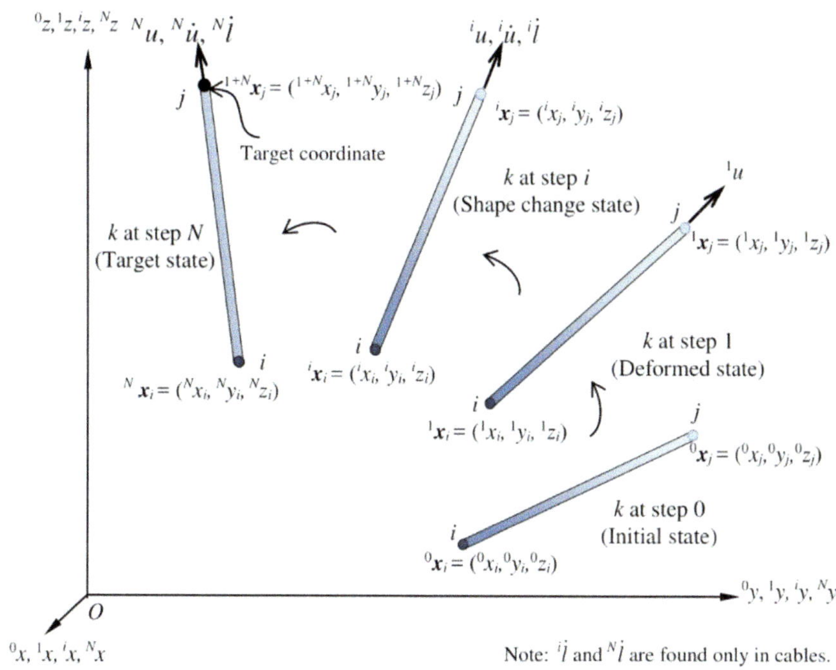

Fig. 8.7 Configurations of element *k* of tensegrity model during shape change analysis

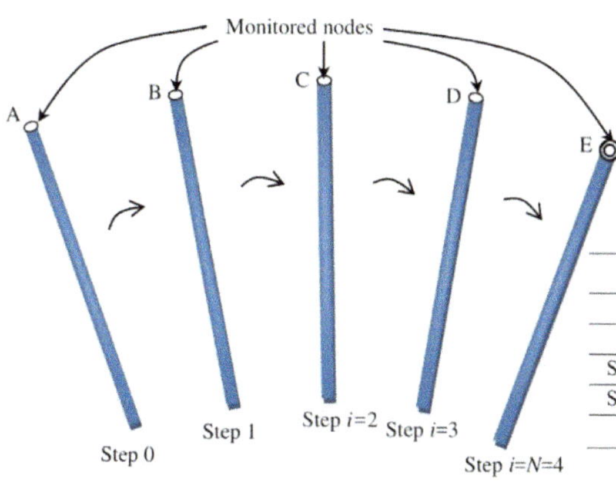

State	Step	${}^{0}x$'	${}^{i}x$'
Initial	0	A	A
Deformed	1	A	B
Shape Change	*i*=2	B	C
Shape Change	*i*=3	C	D
Target	*i*=*N*=4	D	E = $\bar{x}$

Fig. 8.8 Example of nodal coordinates of monitored nodes at previous and current steps

$$ {}^{t}\dot{\mathbf{F}} = {}^{t}\dot{\mathbf{B}}{}^{t}\mathbf{n} + {}^{t}\mathbf{B}{}^{t}\dot{\mathbf{n}} \tag{8.11} $$

Noted that $\mathbf{B}$ is a $n_u \times m$ matrix of directional cosines. The vector of directional cosine λ for a member of a biotensegrity model at incremental step t could be written as follows:

$$ {}^{t}\boldsymbol{\lambda} = \frac{\left({}^{t-1}\boldsymbol{x}_j - {}^{t-1}\boldsymbol{x}_i\right)}{{}^{t-1}l + {}^{t-1}u} \tag{8.12} $$

where l, u, x_i and x_j are the initial geometrical inputs particularly the length after forced elongation, elastic elongation, nodal coordinate at nodes i and j, obtained from the previous step $t-1$.

The incremental equation for each member could be expressed in matrix form as follows:

$$ \begin{Bmatrix} {}^{t}\dot{\mathbf{f}}_i \\ {}^{t}\dot{\mathbf{f}}_j \end{Bmatrix} = \begin{bmatrix} -{}^{t}\dot{\lambda} \\ {}^{t}\dot{\lambda} \end{bmatrix} {}^{t}\mathbf{n} + \begin{bmatrix} -{}^{t}\lambda \\ {}^{t}\lambda \end{bmatrix} {}^{t}\dot{\mathbf{n}} \tag{8.13} $$

where the $\mathbf{f}_i$, $\mathbf{f}_j$ are vectors of nodal force at the end i and j of the member.

The first term on the right-hand side of Eq. (8.13) yields

$$ \begin{bmatrix} -{}^{t}\dot{\lambda} \\ {}^{t}\dot{\lambda} \end{bmatrix} {}^{t}n = {}^{t}\mathbf{k}_g \begin{Bmatrix} {}^{t}x_i \\ {}^{t}x_j \end{Bmatrix} \tag{8.14} $$

where ${}^{t}\mathbf{k}_g$ is the 6×6 geometrical stiffness matrix, which is defined as follows:

$$ {}^{t}\mathbf{k}_g = \frac{{}^{t-1}n}{{}^{t-1}l + {}^{t-1}u} \begin{bmatrix} \left[\mathbf{I}_3 - {}^{t}\boldsymbol{\lambda}{}^{t}\boldsymbol{\lambda}^{\mathrm{T}}\right] & -\left[\mathbf{I}_3 - {}^{t}\boldsymbol{\lambda}{}^{t}\boldsymbol{\lambda}^{\mathrm{T}}\right] \\ -\left[\mathbf{I}_3 - {}^{t}\boldsymbol{\lambda}{}^{t}\boldsymbol{\lambda}^{\mathrm{T}}\right] & \left[\mathbf{I}_3 - {}^{t}\boldsymbol{\lambda}{}^{t}\boldsymbol{\lambda}^{\mathrm{T}}\right] \end{bmatrix} \tag{8.15} $$

where I_3 is a 3×3 identity matrix. The second term on the right-hand side of Eq. (8.13) could be expressed as

$$ \begin{Bmatrix} -{}^{t}\lambda \\ {}^{t}\lambda \end{Bmatrix} {}^{t}\dot{n} = {}^{t}\mathbf{k}_e \begin{Bmatrix} {}^{t}x_i \\ {}^{t}x_j \end{Bmatrix} - \begin{Bmatrix} -{}^{t}\lambda \\ {}^{t}\lambda \end{Bmatrix} \frac{EA}{{}^{t-1}l} \left(1 + \frac{{}^{t-1}u}{{}^{t-1}l}\right) {}^{t}\dot{l} \tag{8.16} $$

where ${}^{t}\mathbf{k}_e$ is the 6×6 linear stiffness matrix which is expressed as

$$ {}^{t}\mathbf{k}_e = \frac{EA}{{}^{t-1}l} \begin{bmatrix} {}^{t}\boldsymbol{\lambda}{}^{t}\boldsymbol{\lambda}^{\mathrm{T}} & -{}^{t}\boldsymbol{\lambda}{}^{t}\boldsymbol{\lambda}^{\mathrm{T}} \\ -{}^{t}\boldsymbol{\lambda}{}^{t}\boldsymbol{\lambda}^{\mathrm{T}} & {}^{t}\boldsymbol{\lambda}{}^{t}\boldsymbol{\lambda}^{\mathrm{T}} \end{bmatrix} \tag{8.17} $$

and is the vector of incremental force elongation of cables. Superimposing Eq. (8.13), and rewritten Eq. (8.11) gives the incremental equation for the entire structure as follows:

$$
\begin{aligned}
{}^t\dot{\mathbf{F}} &= {}^t\mathbf{K}{}^t\dot{\mathbf{x}} - {}^t\mathbf{B}{}^t\mathbf{C}_L{}^t\dot{\mathbf{l}} \\
{}^t\mathbf{K} &= \sum_{\text{member}} \left({}^t\mathbf{k}_e + {}^t\mathbf{k}_g\right)
\end{aligned}
\tag{8.18}
$$

where ${}^t\boldsymbol{K}$ is the structure tangent stiffness matrix. ${}^t\mathbf{C}_L$ is a $m \times m$ matrix which could be further expressed as follows:

$$
{}^t\mathbf{C}_L = \begin{bmatrix} \frac{E_1A_1}{{}^{t-1}l_1} & & & \\ & \frac{E_2A_2}{{}^{t-1}l_2} & & \\ & & \ddots & \\ & & & \frac{E_mA_m}{{}^{t-1}l_m} \end{bmatrix} + \begin{bmatrix} \frac{E_1A_1}{{}^{t-1}l_1}\left(\frac{{}^{t-1}u_1}{{}^{t-1}l_1}\right) & & & \\ & \frac{E_2A_2}{{}^{t-1}l_2}\left(\frac{{}^{t-1}u_2}{{}^{t-1}l_2}\right) & & \\ & & \ddots & \\ & & & \frac{E_mA_m}{{}^{t-1}l_m}\left(\frac{{}^{t-1}u_m}{{}^{t-1}l_m}\right) \end{bmatrix}
\tag{8.19}
$$

Since the tensegrity under feasible prestress is a stable structure, the inverse of exists. The vector of incremental coordinates for the entire structure could be obtained by solving Eq. (8.18) under constant-loading:

$$
{}^t\dot{\mathbf{x}}_G = {}^t\mathbf{K}^{-1t}\mathbf{B}{}^t\mathbf{C}_L{}^t\dot{\mathbf{l}}
\tag{8.20}
$$

While the vector of incremental axial force could be expressed as follows:

$$
\begin{aligned}
{}^t\dot{\mathbf{n}} &= {}^t\mathbf{D}_2{}^t\dot{\mathbf{l}} \\
{}^t\mathbf{D}_2 &= \left({}^t\mathbf{C}_1{}^t\mathbf{B}^{Tt}\mathbf{K}^{-1t}\mathbf{B} - \mathbf{I}_m\right){}^t\mathbf{C}_L
\end{aligned}
\tag{8.21}
$$

Optimisation by Sequential Quadratic Programming. The shape change strategy is planned for a set of identified monitored nodes to advance from their initial position to the target coordinates. This can be achieved by minimising the following objective function with every step:

$$
\min f(\boldsymbol{x}) = \left|{}^t\boldsymbol{x}' - \overline{\boldsymbol{x}}\right|
\tag{8.22}
$$

where $\overline{\boldsymbol{x}}$ is prescribed target coordinates, and ${}^t\,\boldsymbol{x}' = {}^{t-1}\,\boldsymbol{x}' + {}^t\dot{\boldsymbol{x}}'$ is the current coordinates, for all the specified monitored nodes at current step t during the shape change analysis. Sequential Quadratic Programming (SQP) method is applied to solve the nonlinear optimisation problem, as shown in Eq. (8.22). The quadratic programming sub-problem can be written as follows:

$$\begin{aligned}&\min_{x\in R^n} f(\boldsymbol{x}) = \boldsymbol{g}^{T\,t}\dot{\mathbf{l}} + \frac{1}{2}{}^{t}\dot{\mathbf{l}}^{\mathrm{T}}\boldsymbol{H}{}^{t}\dot{\mathbf{l}}\\&\text{subject to } \boldsymbol{A}_2{}^{t}\dot{\mathbf{l}} \geq \boldsymbol{b}_2\\&\text{where } \boldsymbol{g} = \boldsymbol{B}_2^{\mathrm{T}}({}^{t-1}\mathbf{x}' - \overline{\mathbf{x}})\\&\boldsymbol{H} = \boldsymbol{B}_2^{\mathrm{T}}\boldsymbol{B}_2\end{aligned} \tag{8.23}$$

${}^{t}\dot{l}$ is a $m_{cf} \times 1$ vector of incremental forced elongation corresponding to the optimisation variable, and $\boldsymbol{H}$ is a positive-definite approximation of the Hessian matrix of the Lagrangian function. m_{cf} is the number of cables, excluding the undeformed cables. Matrix $\mathbf{A}_2$ and vector $\mathbf{b}_2$ are the inequality constraints corresponding to the limitation in allowable axial forces and forced elongation, respectively, as follows:

$$\boldsymbol{A}_2 = \begin{Bmatrix} \boldsymbol{D}_2 \\ \cdots \\ -\boldsymbol{D}_2 \\ \cdots \\ \boldsymbol{I}_{mcf} \\ \cdots \\ -\boldsymbol{I}_{mcf} \end{Bmatrix}; \boldsymbol{b}_2 = \begin{Bmatrix} \boldsymbol{n}_L - {}^{t-1}\boldsymbol{n} \\ \cdots \\ -(\boldsymbol{n}_U - {}^{t-1}\boldsymbol{n}) \\ \cdots \\ -d_{cl} \\ \cdots \\ -d_{cl} \end{Bmatrix} \tag{8.24}$$

where $\mathbf{A}_2$ is a $(2\,m + 2m_{cf}) \times m_{\mathrm{cf}}$ matrix and b_2 is a $(2\,m + 2m_{cf}) \times 1$ vector. $\boldsymbol{n}_L, {}^{t-1}\boldsymbol{n}, \boldsymbol{n}_U$ are vectors of lower limit of axial force, axial force at the previous step, and upper limit of axial force, respectively. The lower and upper axial force limit for cables and struts complies with Eq. 8.10. d_{cl} (with $d_{cl} > 0$) is the limit for length imposed on each variable forced elongation ${}^{t}\dot{l}$. The obtained converged ${}^{t}\dot{\mathbf{l}}$ is then used in each incremental step of shape change analysis to determine the updated incremental nodal coordinates in Eq. (8.20) and incremental axial forces in Eq. (8.21). Figure 8.9 shows examples of shape change cases in multi-directional modes. Numerical application of the shape change strategy for regular tensegrity prisms and spine biotensegrity models in different directional modes can be found in Oh et al., (2019, 2022).

8.6 Shape Change Considering Obstacle Avoidance

Potential method is one of the strategies to simulate flexible motion in biotensegrity models, enabling them to mimic human movement by reaching target positions while avoiding obstacles. The potential method defines a potential field function that guides the structure's movement along the path of steepest descent. As the structure approaches an obstacle, the potential field value asymptotes to positive infinity, indicating increased difficulty in navigation. Conversely, as the structure nears the target, the function value approaches negative infinity, indicating a successful path. The

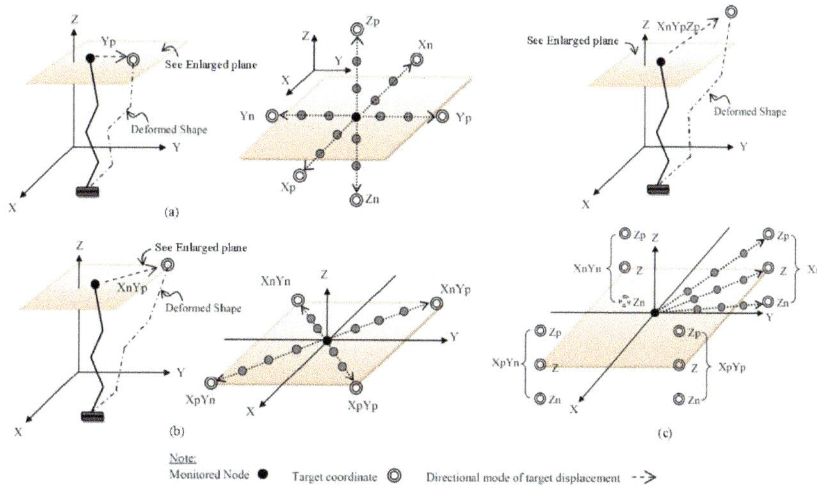

Fig. 8.9 Shape change cases: **a.** uni-; **b.** bi- and **c.** tri-directional modes (Oh et al., 2022)

structure follows the negative gradient of the potential field, effectively navigating away from obstacles and toward the target.

Let us now set the potential field as follows:

$$P({}^t\boldsymbol{x}') = \frac{1}{|{}^t\boldsymbol{x}' - \boldsymbol{x}_o| - r} - \frac{w_T}{|{}^t\boldsymbol{x}' - \overline{\mathbf{x}}|} \tag{8.25}$$

in which r is the radius of an obstacle, w_T is the weight coefficient, ${}^t\boldsymbol{x}'$ is the coordinates of the evaluation node for the target, $\overline{\mathbf{x}}$ is the target coordinate, $\boldsymbol{x}_o$ is the centre of the obstacle (Fig. 8.10a). Say the obstacle is a cylinder. Figure 8.10a illustrates the potential field in Eq. (8.25). The function values are $+\infty$ on the obstacle surface and $-\infty$ at the target coordinate $\overline{\mathbf{x}}$.

The weight coefficient w_T for the target has a significant impact on the avoidance path. Figure 8.10b is a test scenario used to set w_T. Figure 8.10c illustrates the relationship between the w_T and the computational steps. When w_T is small, the number of computational steps is high; however, as w_T increases, the computational steps decrease. For $w_T \geq 60$, the computational steps become nearly constant, and the potential field (orange line) satisfies $P({}^t\boldsymbol{x}') \leq -1$. Therefore, we adopt $P({}^t\boldsymbol{x}') = -1$ at the initial state, as shown in Fig. 8.10.

Nodes other than the evaluation node or members may come into contact with obstacles (Fig. 8.11). The potential function in Eq. (8.26) is applied to address this issue. This function has the same form as the first term in Eq. (8.25).

$$P_A(\boldsymbol{x}_A) = \frac{1}{|\boldsymbol{x}_A - \boldsymbol{x}_o| - r} \tag{8.26}$$

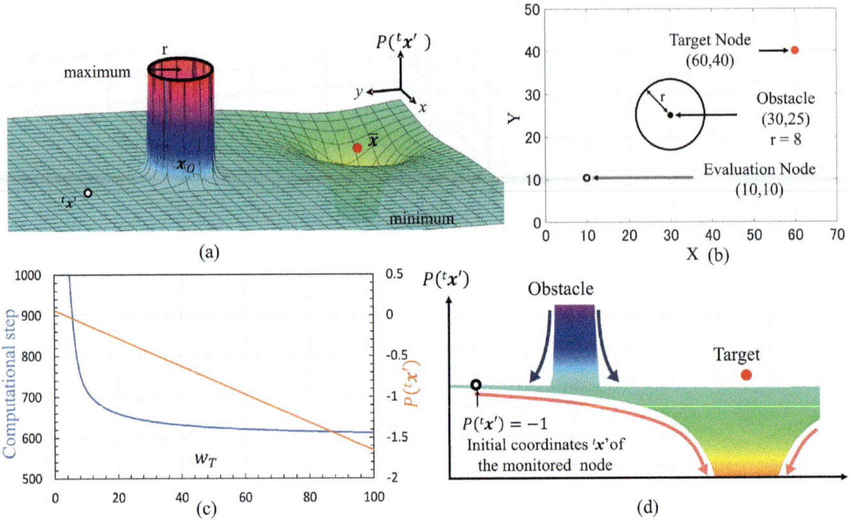

Fig. 8.10 **a** Potential field, **b** Test problem, **c** Relationship between step, potential and w_T, **d** setting of w_T

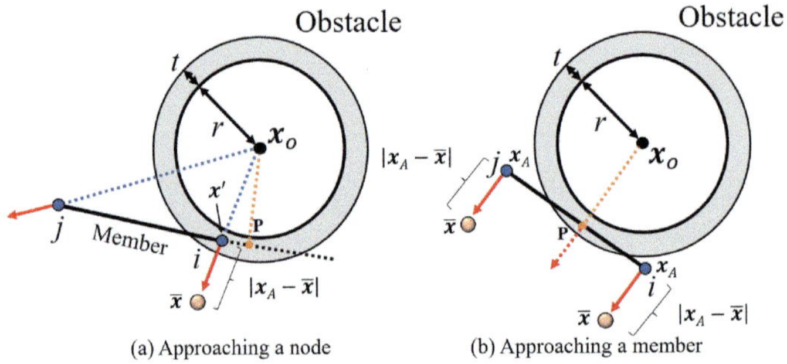

Fig. 8.11 Selection of avoidance nodes

in which x_A is nodal coordinates that avoid contact.

When nodes or members approach the vicinity of the obstacle, they are moved away from it (Fig. 8.11). The direction of movement is given by $\partial(x_A)/\partial x_A < 0$. In Fig. 8.11, the symbol t represents the margin for detecting when an obstacle is nearby. Point **P** is the intersection of the member line with a perpendicular line drawn from the center of the obstacle to the member. The approach depends on whether point **P** is located between the ends i and j of the member. If the i-end approaches the obstacle (Fig. 8.11a), the avoidance node $x_A = x_i$. If a node on a member, excluding end nodes, approaches the obstacle (Fig. 8.11b), $x_A = x_i$ and x_j.

Consider two strategies for avoiding obstacles. The first strategy prioritises the avoidance behaviour of the node closest to the obstacle, even if it means sacrificing the evaluation node's ability to reach the target. The second strategy involves setting avoidance nodes on all nodes near the obstacle while ensuring the evaluation node reaches the target. The second strategy provides smoother control and was therefore chosen for the numerical application.

Optimisation problem for shape control. In this study, temporary target coordinates are set sequentially, and the shape is controlled to reach the target position while avoiding obstacles. The temporary target coordinates are set in the direction where the potential field value decreases, i.e., $\partial({}^t\boldsymbol{x}')/\partial{}^t\boldsymbol{x}' < 0$, $\partial P(\boldsymbol{x}_A)/\partial \boldsymbol{x}_A < 0$, with a finite small distance $|\boldsymbol{x}_A - \overline{\mathbf{x}}|$. The positions of the evaluation node (Fig. 8.12a) for the target and the avoidance nodes (Fig. 8.12b) are determined as the solution to the following optimisation problem. Collisions between members are not considered in this formulation.

Let the coordinates of control nodes and temporary target node at a time ${}^t\boldsymbol{x} = x'(t), \overline{\boldsymbol{x}} - \overline{\boldsymbol{x}}(t)$, respectively. Set the following objective function.

$$f = \frac{1}{2}\left({}^t\boldsymbol{x}' - \overline{\boldsymbol{x}}\right)^T\left({}^t\boldsymbol{x}' - \overline{\boldsymbol{x}}\right) \geq 0 \tag{8.27}$$

The coordinate ${}^t\boldsymbol{x}'$ posterior to time increment Δt is written as

$${}^t\boldsymbol{x}' = {}^{t-1}\boldsymbol{x}'_0 + \dot{\boldsymbol{x}}' \tag{8.28}$$

in which ${}^{t-1}x'(t = t_0)$ is current coordinate, x' is nodal displacement. Substituting Eq. (8.28) into Eq. (8.27) and omitting the constant, we obtain

$$f = \frac{1}{2}\dot{\boldsymbol{x}}'^T\dot{\boldsymbol{x}}' + \left({}^{t-1}x' - \overline{\boldsymbol{x}}\right)^T\dot{\boldsymbol{x}}' \tag{8.29}$$

The relation between the control node coordinates $\acute{x}$ and free node coordinates $\boldsymbol{x}$ can be expressed as

$$\acute{\boldsymbol{x}} = \boldsymbol{B}\boldsymbol{x} \tag{8.30}$$

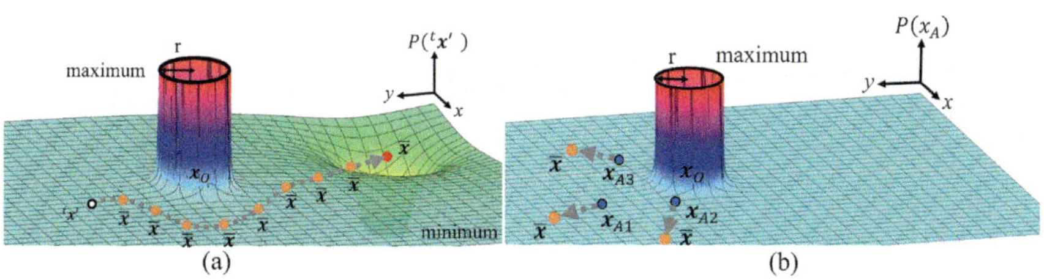

Fig. 8.12 Setting of a temporary target for **a** evaluation node, **b** avoiding node

in which $\boldsymbol{B}$ is a matrix that corresponds to the coordinates of the control points and is constant during time increment Δt. The row size of matrix $\boldsymbol{B}$ is 3 for the first strategy. Because the control node is only an evaluation node for the target. In the case of the second strategy, the row size of matrix $\boldsymbol{B}$ is $3 + 2 \times n_c$ (n_c is the number of avoidance nodes except the evaluation node).

Substitution of Eq. (8.30) into Eq. (8.29) yields

$$f = \frac{1}{2}\dot{\boldsymbol{x}}^T\boldsymbol{B}^T\boldsymbol{B}\dot{\boldsymbol{x}} + \left({}^{t-1}\boldsymbol{x}' - \overline{\boldsymbol{x}}\right)^T\boldsymbol{B}\dot{\boldsymbol{x}} \tag{8.31}$$

By substituting Eq. (8.20) into Eq. (8.31), the shape control reduces to the following sequential quadratic programming problem for $\dot{\mathbf{L}}$.

$$\begin{aligned} &\text{Minimize} f = \frac{1}{2}\dot{\mathbf{L}}^T\mathbf{H}\dot{\mathbf{L}} + \boldsymbol{D}^T\dot{\mathbf{L}} \\ &\text{s.t.} \boldsymbol{n}_L \le \boldsymbol{n} \le \boldsymbol{n}_u \left|\dot{\mathbf{L}}\right| \le \dot{\mathbf{L}}_{\text{lim}} \end{aligned} \tag{8.32}$$

in which $\boldsymbol{n}_L$,$\boldsymbol{n}_U$ in Eq. (8.32) are the lower and upper bound of axial force. $\dot{\mathbf{L}}_{\text{lim}}$ is the move limit (Fig. 8.13).

Numerical application. The shape change methods from Sect. 8.6 and Sect. 8.7 were applied to a spine biotensegrity model SB2 from Oh et al., (2020) (Fig. 8.114a). Table 8.1 summarises the analysis cases, and Table 8.2 shows the section and material properties. Case 1 employs the shape change method without the obstacle avoidance feature, while Case 2 and Case 3 are analysed with the feature. From the viewpoint of biomechanics, when humans change their posture, muscles expand and contract, joints deform, and bones hardly deform. Case 1 and Case 2 allow forced elongation in cables, while Case 3 considers cables and struts. SB2 owns three self-equilibrium stress modes, pinned supports at the bottom three nodes and subjected to vertical

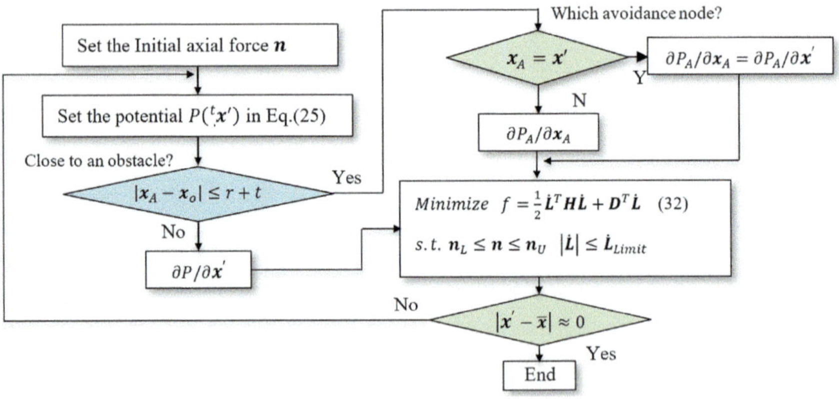

Fig. 8.13 Flow for the proposed shape control

force of 0.5N at all nodes. Figure 8.14b illustrates the positions of the evaluation node, target point, and obstacle, which applied to all the cases. The obstacle is a cylinder, whose radius r is 70 mm. For Case 2 and Case 3, the margin t for detecting the obstacle is 10 mm. The weight coefficient, $w_{t,}$ in Eq. (8.25) is 254. The value is defined from P(^{t}x') = –1 at the initial position. The move limit $\dot{\mathbf{L}}_{\text{lim}} = 0.1$ mm in Eq. (8.32), During each incremental step (Fig. 8.12), the distance from the current control point to the temporary target point $\left|{}^{t}\boldsymbol{x}' - \overline{\mathbf{x}}\right| = 1$ mm. Self-collision between members is not considered in all cases.

Figures 18.15, 18.16 and 18.17 show the shape change of the SB2 model for Case 1, Case 2 and Case 3, respectively. In Case 1, the SB2 model advances directly to the target point by flexion with an obvious bend at the thoracic region and AP2 in Fig. 8.2. Without much changes in the overall width and only a few efforts of the cable's forced elongation at the cervical and thoracic, the model passed through the obstacle and reached the target in fewer computational steps (15 steps). No side twist is observed. The model avoids obstacles in Case 2 and Case 3. Controlling both struts and cable lengths resulted in fewer total computational steps (Case 3 = 708 steps) than controlling only the cable length (Case 2 1026 steps). In both cases, the evaluation node begins to approach the direction of the target point by tilting the upper thoracic region (refer to the dotted frame in Figs. 8.16a and 8.17a). When the model is close to the obstacle, the lower thoracic or lumbar region (refer dotted frame in Figs. 8.16b and 8.17b) expands laterally to evade the obstacle. The model starts to bend over the periphery of the obstacle in clockwise mode, although very close, but without touching the obstacle, and continues to tilt upward (refer to the dotted frame in Figs. 8.16c and 8.17c). Finally, the cervical region extends (refer to the dotted frame in Figs. 8.16d and 8.17d), and the evaluation node reaches the target point. The final configuration of SB2 for Case 2 demonstrates more expansions and contractions compared to Case 3 due to the constraints of strut length.

Figure 8.18 shows the distance to the target point $\left|{}^{t}\boldsymbol{x}' - \overline{\mathbf{x}}\right|$ and the value P(^{t}x') of the potential field in Eq. (8.25) for all cases. The curve for distance for Case 1 is linear compared to Case 2 and Case 3, which decrease monotonically. For Case 3,

Table 8.1 Shape change cases

Case	Obstacle avoidance	Controls	Shape change method
Case 1	Not considered	Cables	Section 8.6, Oh et al., (2022)
Case 2	Considered	Cables	Section 8.7, Hukumori et al., (2022)
Case 3	Considered	Struts and cables	Section 8.7, Hukumori et al., (2022)

Table 8.2 Inputs for spine biotensegrity model

Member	Number of members	Young modulus (GPa)	Cross-sectional area (mm^2)	Yield stress (MPa)
Strut	12	200	50.3	250
Cable	57	200	3.14	250

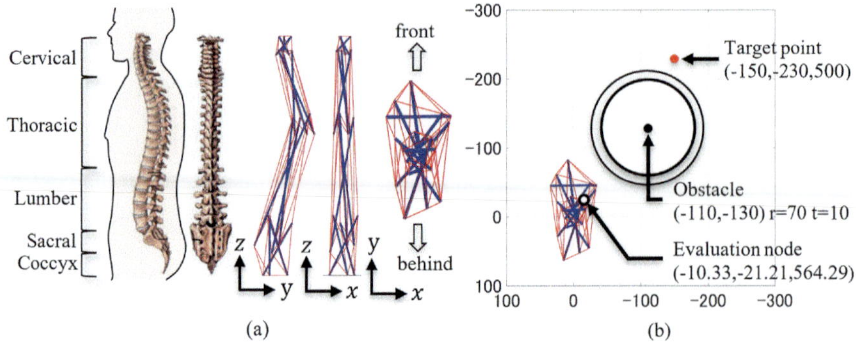

Fig. 8.14 Inputs for spine biotensegrity model **a** Human spine and SB2 model (Oh et al., 2020), **b** Obstacle, target node, and evaluation node unit (mm)

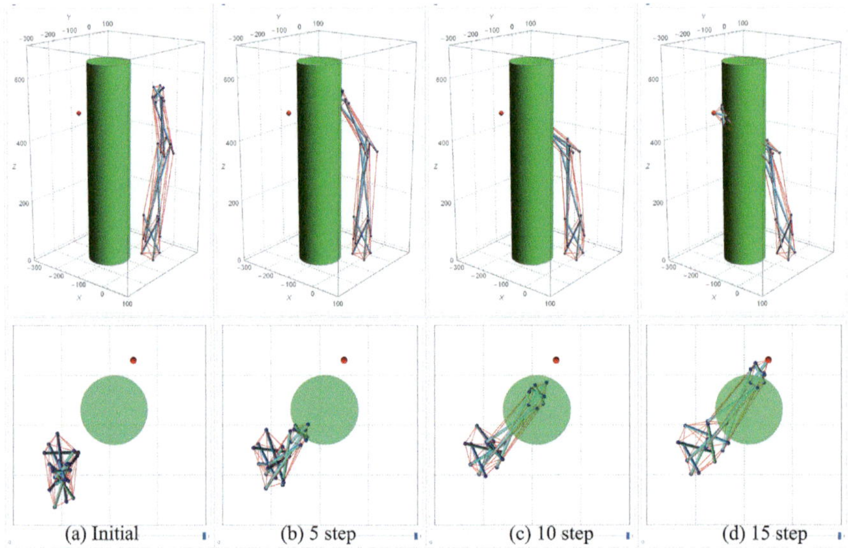

Fig. 8.15 Shape change for case 1 (controls cable length)

approximately at step 472, the evaluation node has started to avoid the obstacle, and the value P($^t x'$) begins to decrease. After step 580, The potential P($^t x'$) falls sharply, which reveals the model extensively avoids obstacles until the evaluation meets the target point. SB2 in Case 2 behave similarly to Case 3; only the obstacle avoidance process occurs at more computational steps due to the limitation of strut length.

Overall, when the biotensegrity model approaches an obstacle, a temporary target point is set in the direction in which the gradient of the potential field decreases, and the avoidance route is continuously searched. This series of processes is formulated as a sequential quadratic programming problem constrained by allowable axial force. The analysis result using only the length of the cable as the control parameter demonstrated the motion of skillfully twisting the entire structure while deforming the joint to avoid obstacles.

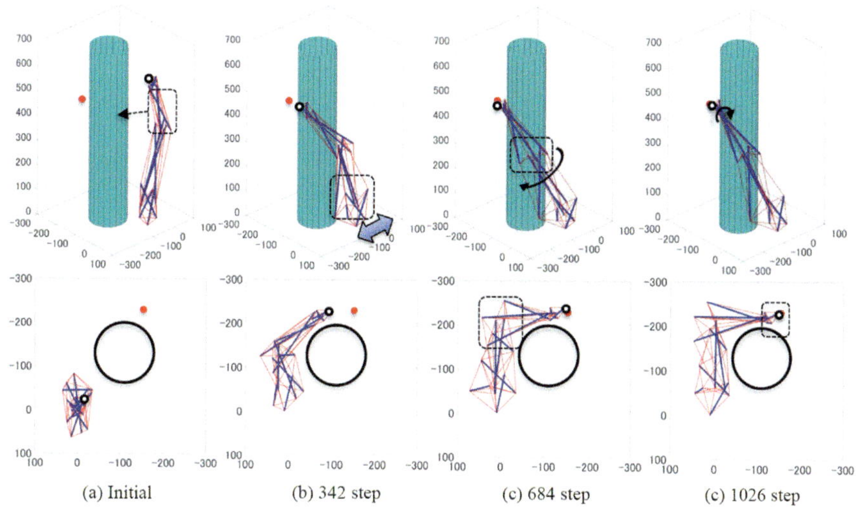

Fig. 8.16 Shape change for case 2 (controls on only cable length)

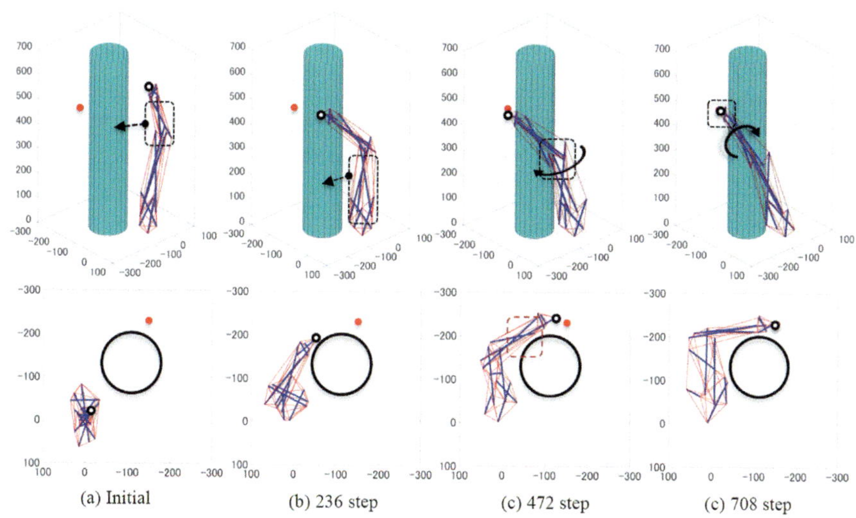

Fig. 8.17 Shape change for case 3 (controls on both strut and cable length)

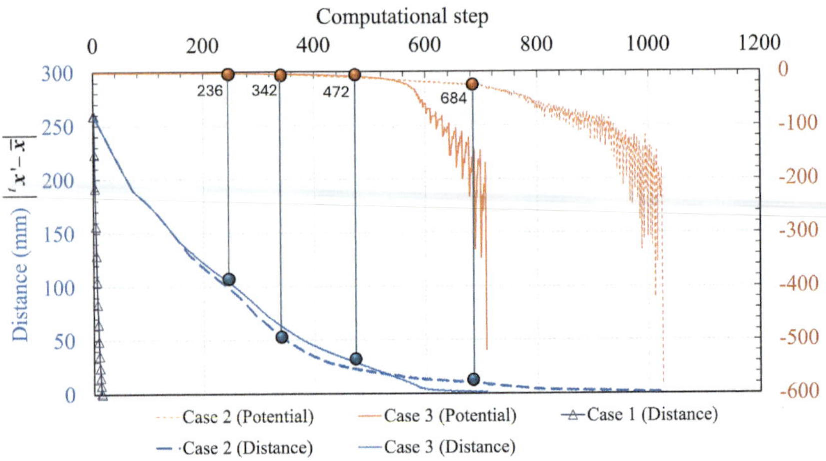

Fig. 8.18 The distance to the target point and potential field

8.7 Concluding Remarks and Future Recommendations

Biotensegrity, an emerging interdisciplinary concept blending biology and tensegrity (tensional integrity), is gaining traction in various fields, including biomechanics, architecture, robotics, and tissue engineering. The shape change control of biotensegrities holds significant potential for further exploration. Future research should focus on the simulation of different obstacles with different shapes and those that surround the biotensegrity model, such as height limits and the existence of walls in the path towards a target. Additionally, developing a shape change algorithm to achieve multiple targets, considering various loadings, and employing sustainable and least energy-efficient shape change modes is essential. The shape change control of biotensegrities could be incorporated into advanced and soft robotics to perform complex tasks, such as inspection, rescue, or biohybrid robots. Utilising the materials from the biology system and the degree of shape change to detect distress or pain in muscles or tissues in simulations of rehabilitation or therapy are recommended for investigation. Virtual reality and simulations of biotensegrity movements could lead to advancements in regenerative medicine, tissue engineering, medical diagnosis and treatment, and rehabilitation engineering. The integration of AI and machine learning with biotensegrity research could accelerate the discovery of new forms and applications.

References

Busscher, I., Ploegmakers, J. J., Verkerke, G. J., & Veldhuizen, A. G. (2010). Comparative anatomical dimensions of the complete human and porcine spine. *European Spine Journal, 19*, 1104–1114.

Edwards, S. A., Wagner, J., & Gräter, F. (2012). Dynamic prestress in a globular protein. *PLoS Computational Biology, 8*(5), e1002509.

Essen, D. C., & v. (1997). A tension-based theory of morphogenesis and compact wiring in the central nervous system. *Nature, 385*(6614), 313–318.

Fraddosio, A., Pavone, G., & Piccioni, M. D. (2021). A novel method for determining the feasible integral self-stress states for tensegrity structures. *Curved and Layered Structures, 8*(1), 70–88.

Hagert, E., & Hagert, C.-G. (2010). Understanding stability of the distal radioulnar joint through an understanding of its anatomy. *Hand Clinics, 26*(4), 459–466.

Hukumori, I., Nishimura, T., Oh, C. L., Choong, K. K., & Kim, J.-Y. (2022). Shape control of tensegrity model mimicking human spine by the potential method. *Proceedings of IASS Annual Symposia.*

Ingber, D. E. (2003a). Mechanical control of tissue morphogenesis during embryological development. *International Journal of Developmental Biology, 50*(2–3), 255–266.

Ingber, D. E. (2010). From cellular mechanotransduction to biologically inspired engineering: 2009 Pritzker award lecture, BMES annual meeting October 10, 2009. *Annals of Biomedical Engineering, 38*, 1148–1161.

Ingber, D. E. (2023). From tensegrity to human organs-on-chips: Implications for mechanobiology and mechanotherapeutics. *Biochemical Journal, 480*(4), 243–257.

Ingber, D. E. (2003b). Tensegrity I. Cell structure and hierarchical systems biology. *Journal of cell science, 116*(7), 1157–1173.

Levin, S. M. (2007). A suspensory system for the sacrum in pelvic mechanics: Biotensegrity. *Movement, Stability and Lumbopelvic Pain, 2*, 229–237.

Levin, S. M. (2002). The tensegrity-truss as a model for spine mechanics: biotensegrity. *Journal of Mechanics in Medicine and Biology*, *2*(03n04), 375–388.

Maina, J. (2007). Spectacularly robust! Tensegrity principle explains the mechanical strength of the avian lung. *Respiratory Physiology & Neurobiology, 155*(1), 1–10.

Oh, C. L., Choong, K. K., Nishimura, T., Kim, J.-Y., & Hassanshahi, O. (2019). Shape change analysis of tensegrity models. *Latin American Journal of Solids and Structures, 16*, e221.

Oh, C. L., Choong, K. K., Nishimura, T., & Kim, J.-Y. (2020). Form-finding of spine inspired biotensegrity model. *Applied Sciences, 10*(18), 6344.

Oh, C. L., Choong, K. K., Nishimura, T., & Kim, J.-Y. (2022). Multi-directional shape change analysis of biotensegrity model mimicking human spine curvature. *Applied Sciences, 12*(5), 2377.

Ohsaki, M., & Zhang, J. (2015). Nonlinear programming approach to form-finding and folding analysis of tensegrity structures using fictitious material properties. *International Journal of Solids and Structures, 69*, 1–10.

Scarr, G. (2008). A model of the cranial vault as a tensegrity structure, and its significance to normal and abnormal cranial development. *International Journal of Osteopathic Medicine, 11*(3), 80–89.

Scarr, G. (2012). A consideration of the elbow as a tensegrity structure. *International Journal of Osteopathic Medicine, 15*(2), 53–65.

Tibert, A., & Pellegrino, S. (2011). Review of form-finding methods for tensegrity structures. *International Journal of Space Structures, 26*(3), 241–255.

Vrtovec, T., Pernuš, F., & Likar, B. (2009). A review of methods for quantitative evaluation of spinal curvature. *European Spine Journal, 18*, 593–607.

Zhou, J.-Y., Chen, W.-J., Zhao, B., Qiu, Z.-Y., & Dong, S.-L. (2015). Distributed indeterminacy evaluation of cable-strut structures: Formulations and applications. *Journal of Zhejiang University-SCIENCE A, 16*, 737–748.

Chapter 9
On the Additive Manufacturing of Tensegrity Systems

Anna Al Sabouni-Zawadzka, Adam Zawadzki, Rana Nazifi Charandabi, Ada Amendola, Howon Lee, and Fernando Fraternali

Abstract This Chapter presents a range of additive manufacturing techniques available for the rapid prototyping of tensegrity systems at various scales. It discusses the opportunities and challenges of these techniques, with the primary challenges arising from difficulties in manufacturing perfectly hinged connections and applying internal prestress. The Chapter concludes by summarizing the key findings from current research on the additive manufacturing of tensegrity systems and highlighting promising directions for future research in this area.

Keywords Tensegrity-inspired metamaterial · Additive manufacturing · Stereolitography · Electron-beam melting · Laboratory testing

---

A. Al Sabouni-Zawadzka (✉)
Faculty of Civil Engineering, Warsaw University of Technology, Al. Armii Ludowej 16, 00-637 Warsaw, Poland
e-mail: anna.zawadzka@pw.edu.pl

A. Zawadzki
Faculty of Automotive and Construction Machinery Engineering, Warsaw University of Technology, Narbutta 84, 02-524 Warsaw, Poland
e-mail: adam.zawadzki@pw.edu.pl

R. Nazifi Charandabi
Department of Information Engineering, Electrical Engineering and Applied Mathematics (DIEM), University of Salerno, Via Giovanni Paolo II 132, 84084 Fisciano, SA, Italy
e-mail: rnazificharandabi@unisa.it

A. Amendola · F. Fraternali
Department of Civil Engineering, University of Salerno, Via Giovanni Paolo II 132, 84084 Fisciano, SA, Italy
e-mail: adaamendola1@unisa.it

F. Fraternali
e-mail: f.fraternali@unisa.it

H. Lee
Department of Mechanical Engineering, Seoul National University, 08826 Gwanak-gu, Seoul, Korea
e-mail: howon.lee@snu.ac

F. Fraternali and J. J. Rimoli (eds.), *Tensegrity Systems*, CISM International Centre for Mechanical Sciences 2, https://doi.org/10.1007/978-3-031-82283-4_9

9.1 Introduction

This book has demonstrated that tensegrity systems offer numerous potential technical applications, ranging from large-scale structures used in civil, mechanical, or aerospace engineering (see, e.g., Rimoli, 2018; Fraternali & de Castro Motta, 2023; Fraternali et al., 2024) to cellular mechanical metamaterials (de Castro Motta et al., 2024; Placidi et al., 2024; Salahshoor et al., 2018; Bauer et al., 2021; Al Sabouni-Zawadzka, 2022).

Depending on the intended scale, various manufacturing techniques can be employed to construct tensegrity structures. Large-scale systems can be assembled using readily available engineering components, such as standard rods and cables. This approach has the advantage of preserving the essential characteristics of a pure tensegrity system: hinged connections, self-stress, and infinitesimal mechanisms.

Fabricating tensegrity metamaterials at smaller scales presents greater challenges. In such cases, additive manufacturing (AM) techniques, particularly 3D printing, are commonly employed. However, a major drawback of these methods is that the resulting systems often cannot be classified as true tensegrity structures, as the connections between members are typically not perfectly hinged, allowing for the transfer of some bending stresses. Consequently, most of the structures discussed in this Chapter are referred to as 'tensegrity-inspired' systems. Another limitation of AM for tensegrity systems is the difficulty in introducing prestressing forces (self-stress) in 3D-printed models. The most common AM techniques used to fabricate tensegrity structures at different scales are reviewed in Sect. 9.2.

Recent studies have focused on developing novel mechanical metamaterials based on 3D-printable tensegrity-inspired systems. Amendola et al. (2015) investigated 3D-printed models of tensegrity prisms and columns, as will be discussed in Sect. 9.3. Pajunen et al. (2019) presented the design of a 3D-printable structure with response characteristics comparable to a pin-jointed tensegrity system. Lee et al. (2020) and Wen et al. (2020) introduced a new method for fabricating smart material-based tensegrity structures by combining 3D printing with sacrificial molding. Vangelatos et al. (2020) demonstrated that multistable tensegrity-based lattices can be manufactured using multiphoton lithography. Pajunen et al. (2021) presented tensegrity-inspired lattice systems with prestrain-induced bandgap tunability. Zeng et al. (2023) introduced a 3D-printable metamaterial with periodic tensegrity-inspired lattice-disc unit cells.

Sections 9.4 and 9.5 focus respectively on the stereolithography-based manufacturing of small-scale and micro-scale tensegrity-inspired systems. The Chapter concludes with remarks on the current state of research and outlines directions for future work in the field of AM for tensegrity systems, as presented in Sect. 9.6.

9.2 Available AM Techniques

Additive manufacturing was introduced in the 1980s, with stereolithography being the first commercial technique for rapid prototyping (Hull, 1984). Since then, AM technology has developed rapidly, and today, many different techniques are available. The advent of 3D printing has been a significant milestone in the research on tensegrity-inspired metamaterials, allowing scientists to study these systems not only analytically or numerically but also experimentally (Amendola et al., 2015; Al Sabouni-Zawadzka et al., 2024a).

AM technology allows us to fabricate physical objects using computer-aided design (CAD) or a digital 3D model. There are three main categories of AM techniques (Surjadi et al., 2019): ink-based, light-based, and powder-based methods.

In ink-based methods, the material is deposited onto a flat substrate by melting a filament and extruding it through a nozzle. Two of the most popular techniques are Fused Deposition Modelling (FDM) (Gao et al., 2020) and Direct Ink Writing (DIW) (Lewis, 2006). FDM is a widely used and accessible technique that employs thermoplastic materials. In DIW, a viscoelastic or viscoplastic ink is used. However, ink-based methods are not well-suited for fabricating tensegrity-based lattices due to the weak bond between consecutive layers, which leads to strong anisotropic properties in the printed materials.

Light-based methods use photopolymerization processes to create very fine features. In these techniques, UV light or a laser is directed into a photopolymer resin. Popular light-based methods that are well-suited for manufacturing tensegrity-inspired metamaterials include Stereolithography (SLA) (Huang, 2020) and Two-Photon Lithography (TPL) (Harinarayana & Shin, 2021), also known as Direct Laser Writing (DLW). TPL takes advantage of the nonlinear dependency of polymerization on the intensity of irradiating light, allowing the fabrication of 3D microstructures with high resolution. Projection micro-stereolithography (PμSL) is an additive fabrication technique based on photopolymerization, capable of generating complex three-dimensional structures ranging from micro- to small-scale. This additive manufacturing system employs a state-of-the-art digital micro-display as a dynamically reconfigurable photomask, which spatially modulates ultraviolet (UV) light emitted from a high-power light-emitting diode (LED). Liquid polymer resin is photochemically converted into a solid 3D microstructure in a layer-by-layer process. PμS also enables the use of swelling gels that expand and contract in response to changes in humidity and temperature (Lee et al., 2012).

In powder-based methods, a laser source fuses powder particles through melting or sintering. The quality of the print depends on the shape, size, and distribution of the powder. Among the most popular powder-based methods are Selective Laser Sintering (SLS) (Gan et al., 2020) and Selective Laser Melting (SLM) (Song et al., 2022). Electron Beam Melting (EBM) is similar to Selective Laser Melting (SLM) in that both processes use a powder bed for 3D printing. However, EBM (Amendola et al., 2015) employs a beam of electrons to melt layers of powdered metal and form the object within a vacuum, while SLM uses a laser beam to melt or sinter the metal powder, solidifying it without the need for a vacuum environment. Both techniques can be used for the commercial fabrication of metallic structures, including metal-based tensegrity-inspired structures and metamaterials.

9.3 EBM and Post-tensioning of Tensegrity Prisms and Columns

The study presented in Amendola et al. (2015) explores the additive manufacturing and experimental testing of 3D models of tensegrity prisms and columns. Structures composed of tensegrity prisms with rigid bases and temporary supports are 3D printed using an Electron Beam Melting (EBM) system, employing the titanium alloy Ti6Al4V. After the additive manufacturing process, the temporary supports are removed, and Spectra cross-strings are manually added to the 3D printed prisms by passing them through holes located on the nodal junctions (Fig. 9.1).

This process aims to 'sew' the 3D printed model together, after which an appropriate level of internal prestress is applied to the structure ('post-tensioning' approach). In more detail, the Spectra strings are threaded through the holes at the joints and secured to the top base using slipknots. Next, the sacrificial supports of the 3D-printed structure are removed, and tension is applied to the strings by anchoring them to the bottom base. During the post-tensioning phase, a level is positioned on the top base

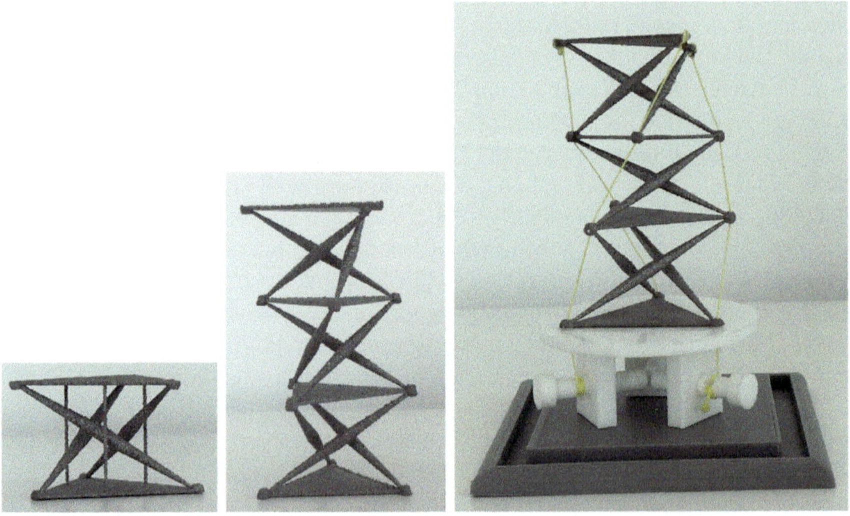

(a) (b) (c)

Fig. 9.1 3D printing of tensegrity prisms (TP) and columns (TC) through electron beam melting in Ti6Al4V titanium alloy: **a** TP with temporary supports; **b** TC after removal of temporary supports; **c** Post-tensioned TC employing Spectra fibers

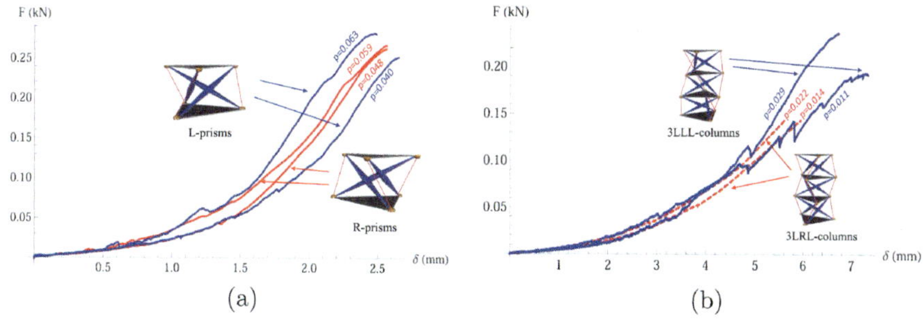

Fig. 9.2 Force F versus displacement δ curves of TPs and TCs additively manufactured with EMB under axial compression loading. A 'left-handed' (L) prism is characterized by the upper base being rotated counterclockwise with respect to the lower base, while a 'right-handed' (R) prism features the upper base rotated clockwise relative to the lower base. The 3LL columns consist of three superimposed L-prisms, whereas the 3LRL column comprises two terminal L-prisms and one central R-prism. The symbol p denotes the prestrain applied to the Spectra strings. Figure reproduced with permission from (Amendola et al., 2015)

to ensure it remains horizontal and to verify that the tensile forces in the strings are evenly distributed. The systems analyzed in Amendola et al. (2015) feature prisms with a height of 22 mm; bi-conical Ti6Al4V struts with a diameter of 0.5 mm at the extremities and a diameter of 3.00 mm at the center; and 0.1 mm Spectra wires (see Amendola et al., 2015 for further details). The experimental results presented in the referenced study show that these structures exhibit an initial stiffening elastic response under large or moderately large axial strains caused by compressive loading (see Fig. 9.2). This is indicated by the tangent stiffness (i.e., the slope of the force-displacement curve) increasing with greater displacements in this region. The stiffening response is followed by a softening phase, where the tangent stiffness decreases as the displacement (δ) increases, approaching the failure point (Fig. 9.2).

This geometrically nonlinear behavior suggests potential applications of tensegrity prisms (TPs) and tensegrity columns (TCs) as innovative mechanical metamaterials and smart devices, whose performance can be effectively tuned by applying suitable states of internal and external prestress, as discussed in the Chapter on solitary waves.

9.4 SLA Manufacturing of Tensegrity-Inspired Structures

This section discusses 3D printing and experimental testing of a series of tensegrity-inspired cells and supercells based on expanded octahedron. All specimens were fabricated using SLA-based 3D printer (Phrozen Sonic Mini 8K) available at the Faculty of Civil Engineering of the Warsaw University of Technology. The first stage of research included designing, additive manufacturing, and testing of single tensegrity-inspired modules (metamaterial cells) made of four different parent materials with the following properties (values declared by the producers of the resins, based on the tests performed on 3D printed dog-bone specimens according to ASTM D 638-14 D63 2014 standard procedures):

- resin no. 1: $E = 2600$ MPa, $f_t = 40$ MPa, $\Delta = 6\%$,
- resin no. 2: $E = 943$ MPa, $f_t = 31$ MPa, $\Delta = 90\%$,
- resin no. 5: $E = 600$ MPa, $f_t = 19$ MPa, $\Delta = 30\%$,
- resin no. 6: $E = 1000$ MPa, $f_t = 55$ MPa, $\Delta = 40\%$,

where: E—Young's modulus, f_t—tensile strength, and Δ—elongation at break.

For the second stage of research, one parent material was chosen—resin no. 6—and expanded octahedron-based supercells were designed, manufactured, and tested. Geometry of the expanded octahedron module and the method for creating supercells is presented in Fig. 9.3. The supercells consist of eight modules arranged in such a way that they form a repetitive unit with three symmetry planes.

Diameters of cables and struts were designed in such a way that the cable-to-strut stiffness ratio: $k = (EA)_{\mathrm{cable}}/(EA)_{\mathrm{strut}} = 0.1$, where E is the elastic modulus of the parent material and A is the cross-sectional area of the member. With this assumption, the diameter of struts was adopted as 3 mm and for cables—0.95 mm.

CAD models of the specimens were prepared in SolidWorks 2014 software, and then transferred to ChiTuBox Slicer in order to add supports and resin profile to prepare input files for the 3D printer. Models of the cell and supercell with the supports are presented in Fig. 9.4. Figure 9.5 shows photographs of the manufacturing process—3D printing of the specimens and post-processing in UV light.

The test series consisted either of seven (single modules) or three (supercells) specimens. Photos of selected specimens—one per each tested parent material and geometrical configuration—are presented in Figs. 9.6 and 9.7a. Symbols used in the specimen designations are following: C—cell, S—supercell, oct—expanded octahedron, {1,2,5,6}—number of the resin, s3—scale 3 cm (single module inscribed into a cube of edge length 3 cm).

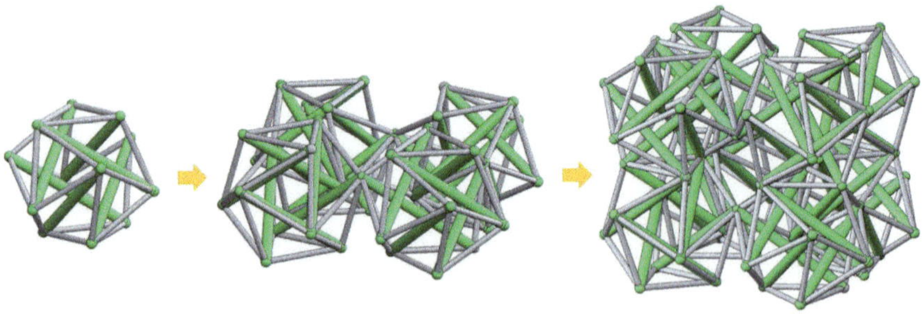

Fig. 9.3 Construction of expanded octahedron lattices

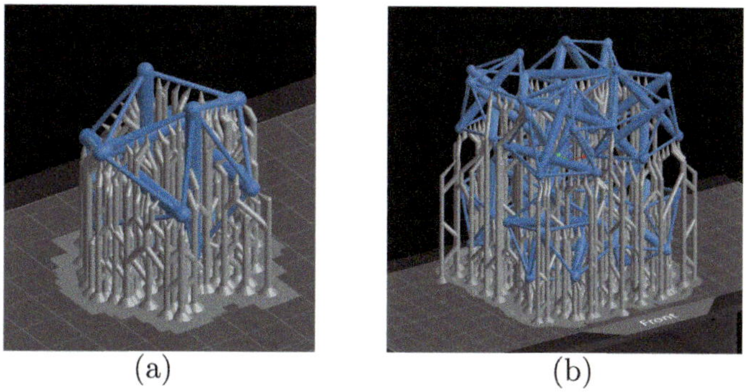

Fig. 9.4 Supported models prepared for 3D printing: **a** cell; **b** supercell

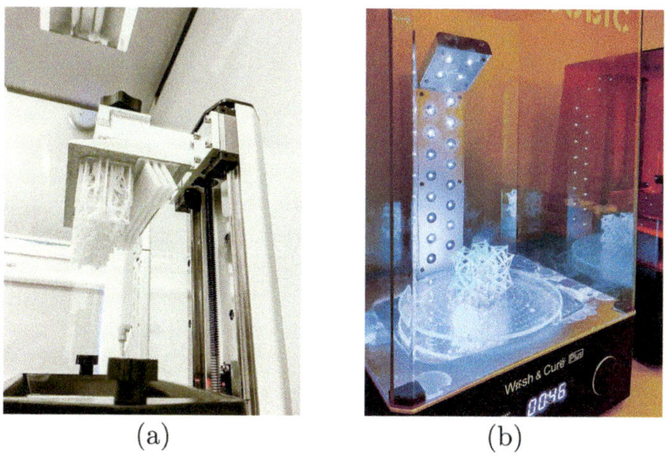

Fig. 9.5 Manufacturing process of the specimens: **a** 3D printing; **b** post-processing in UV light

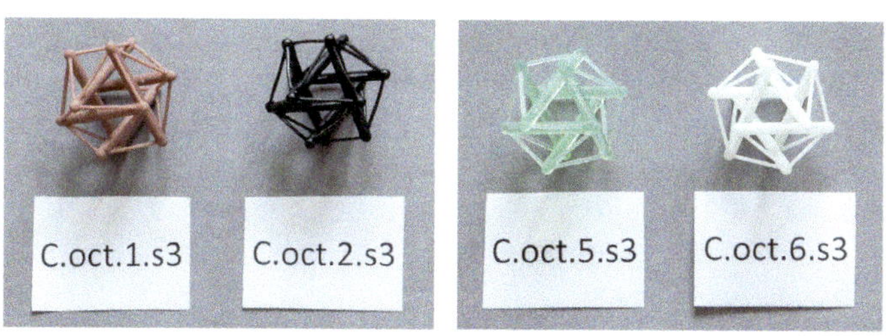

Fig. 9.6 3D-printed expanded octahedron modules

All 3D-printed specimens were subjected to uniaxial compression tests in the universal testing machine INSTRON 5567, with the measurement accuracy of 0.5 N. Photo of the test stand with the lattice under compression is presented in Fig. 9.7b. A

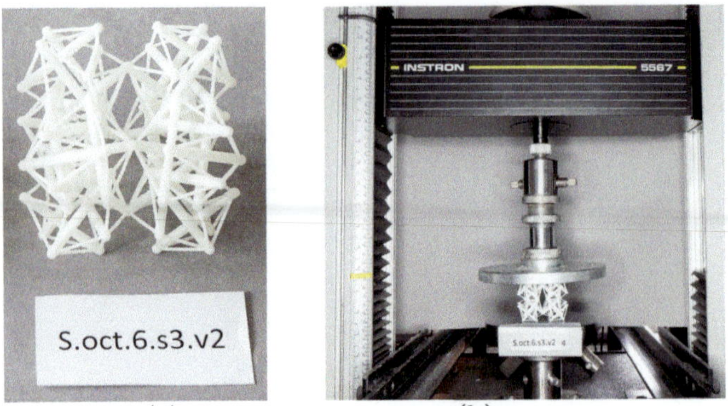

(a) (b)

Fig. 9.7 Expanded octahedron lattice: **a** 3D-printed specimen; **b** compression test

displacement control method with a velocity of 2 mm/min was applied. The results in the form of force corresponding to the displacement in time were obtained directly from the universal testing machine.

Results of the tests performed on single cells are depicted in Fig. 9.8. Each curve was determined as a mean of seven measurements and for each test series a standard deviation was calculated and marked with vertical lines (Fig. 9.8). Moreover, values of three basic parameters (Young's modulus, tensile strength, and elongation and break) of the resins are given next to each curve to facilitate interpretation of the results.

The obtained shapes of the curves are directly related to the properties of the parent materials. It can be noticed that the specimens made of materials with high values of the elongation at break (resins nos. 2, 5, and 6) exhibit a strongly non-linear behavior, as opposed to specimens made of resin no. 1, which got destroyed abruptly, reaching relatively small values of displacements. Another important observation is that the higher the Young's modulus of the resin, the stiffer tensegrity-inspired structure is obtained. The results in Fig. 9.8 show that the specimen made from resin no. 1 initially exhibits an elastically stiffening response, which next transitions into a linear regime, for increasing values of the axial displacement. In contrast, the force-displacement curves of the specimens made from resins nos. 2, 5, and 6 display a stiffening response, followed by a softening phase that ends in a plateau near the failure point.

Averaged results of compression tests conducted on three supercell specimens, with standard deviation marked, are shown in Fig. 9.9. The behavior of the supercell is similar to the observed behavior of the single module made of corresponding resin. It exhibits a stiffening response, followed by a softening phase that ends in a plateau near the failure point.

Based on the obtained test results and the similarity bewteen the behavior of single cells and supercells, it can be concluded that the metamaterial built from the proposed

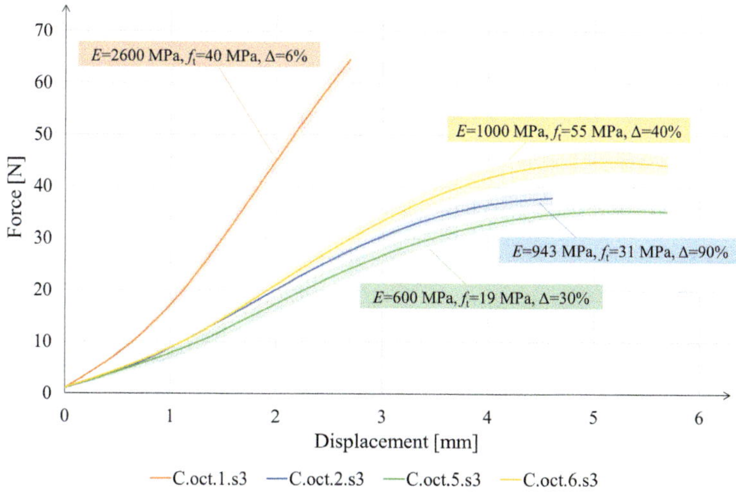

Fig. 9.8 Expanded octahedron modules—compression test results

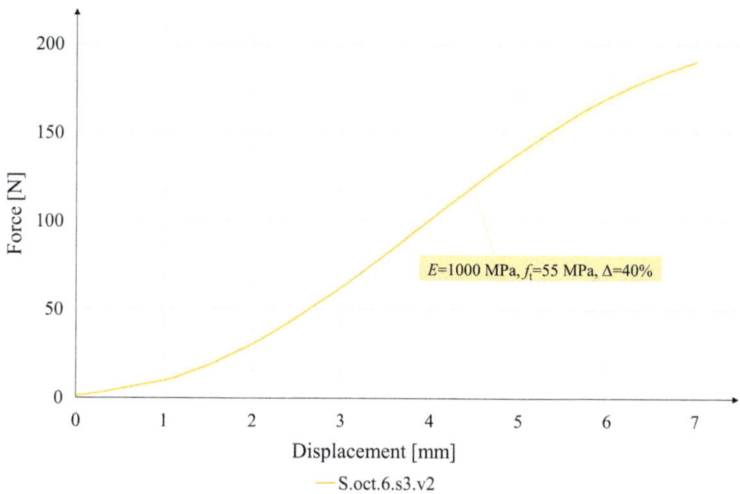

Fig. 9.9 Expanded octahedron supercell—compression test results

supercells would exhibit similar mechanical characteristics to the supercells. This finding confirms results of theoretical study (discrete and continuum approaches) on tensegrity metamaterials (Al Sabouni-Zawadzka, 2022)—the observation that the tensegrity lattice of any size behaves in the same way as its repetitive unit.

9.5 PμSL Manufacturing of Micro-scale T3 Prisms

In recent years, research groups from the Department of Civil Engineering at the University of Salerno and the Department of Mechanical Engineering at Seoul National University have successfully fabricated micro-scale T3 prisms using a custom-built projection micro-stereolithography (PμSL) system (Han et al., 2019; Fraternali & Lee, 2024), as shown in Fig. 9.10.

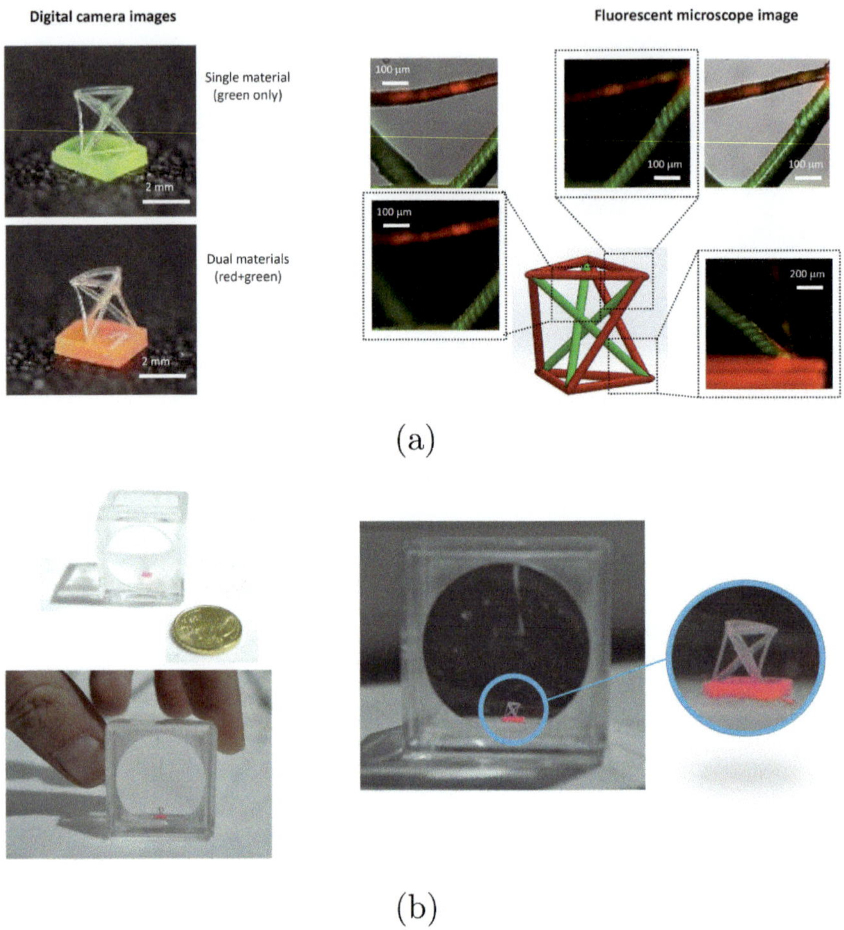

Fig. 9.10 Images of PμSL-printed micro-scale T3 prisms: **a** Single- and double-material specimens. The diameter of cable and strut members is 110 μm; the height and the width of prism is 1.8 mm. Both material A and material B are PEGDA 250 (200 mL). With reference to the double-material specimen, material A (red) contains Rhodamine B (1 mM) as the dye (0.096 g) and is used for the cables, while material B (green) contains $DiOC_2$ (1 mM) as the dye (0.092 g) and is used for the bars. The single-material specimen is instead made solely of material B. **b** Close-up showing the millimeter-scale size of a PμSL-printed sample. Partially reproduced with permission from Fig. 9.4b of Han et al. (2019)

These prisms utilized swelling materials for the tensile members (Lee et al., 2012), which contract upon drying to generate internal prestress. The PμSL technique can effectively employ active hydrogels, such as poly(ethylene glycol) diacrylate (PEGDA), hexanediol diacrylate (HDDA), and poly(N-isopropylacrylamide) (PNIPAAm) (Zheng et al., 2012; Han et al., 2019). Global precompression can also be applied using magnets, external cables tensioned by pushing against the lattice with anchoring plates, or other tools. This provides additional prestress in case the internal prestress is insufficient.

9.6 Concluding Remarks

This Chapter has reviewed a range of techniques available in the literature for the additive manufacturing (AM) of tensegrity structures, including electron-beam melting (Amendola et al., 2015), stereolithography (SLA) (Al Sabouni-Zawadzka et al., 2024b), and projection micro-stereolithography (Fraternali & Lee, 2024). Other notable techniques, such as the use of 3D-printed shape memory polymers that change shape when heated after deposition (Liu et al., 2017), have also been proposed, though they fall outside the scope of this study.

It has been observed that most of the 3D-printed tensegrity structures examined in this study exhibit a nonlinear constitutive response characterized by an initial elastically stiffening behavior under compressive loading, which eventually transitions into a softening response as displacement increases. This geometrically nonlinear behavior can be leveraged to develop innovative nonlinear mechanical metamaterials with a tensegrity architecture, as explored in other chapters of this book (see, for example, the chapter dedicated to the solitary wave dynamics of tensegrity lattices).

One issue noted is that 3D-printed structures often feature joints that are not perfectly hinged and cables that exhibit non-negligible bending stiffness, unlike in true tensegrity systems. Some studies have addressed this limitation by incorporating non-3D-printed cables with additively manufactured bars (Amendola et al., 2015; Liu et al., 2017). However, fully 3D-printing a structure that conforms to the strict definition of a tensegrity system-an assembly of ‘strings’ or ‘cables’ that carry only tensile forces, ‘struts’ that carry only compressive forces, and ‘bars’ that can carry either compressive or tensile forces, as defined in the Preface of this book-remains a challenge.

This challenge represents an important direction for future research in the AM of tensegrity structures. Other promising areas for exploration include the use of materials with different coefficients of thermal expansion to induce internal prestress in tensegrity lattice prototypes; the AM of microlattices using PμSL with swelling materials for the tensile members, which contract upon drying to generate internal prestress; and the fabrication of large-scale tensegrities through robotic arm-based AM techniques (Askari et al., 2020).

References

ASTM D 638-14: Standard test method for tensile properties of plastics (pp. 1–17) (2014)

Al Sabouni-Zawadzka, A. (2022). *High performance tensegrity-inspired metamaterials and structures*. CRC Press.

Al Sabouni-Zawadzka, A., Gilewski, W., Zawadzki, A. (2024a) Experimental investigations on mechanical propertiesof 3d-printed tensegrity-inspired metamaterialsbased on 4-strut simplex module. *Archives of Civil Engineering, 70*(3), 343–357.

Al Sabouni-Zawadzka, A., Gilewski W., Nazifi Charandabi, R., & Zawadzki, A. (2024b). Stability of tensegrity-inspired structures fabricated through additive manufacturing. *Composite Structures, 345*, 118377.

Amendola, A., Hernández-Nava, E., Goodall, R., Todd, I., Skelton, R. E., & Fraternali, F. (2015). On the additive manufacturing, post-tensioning and testing of bi-material tensegrity structures. *Composite Structures, 131*, 66–71.

Askari, M., Hutchins, D. A., Thomas, P. J., Astolfi, L., Watson, R. L., Abdi, M., Ricci, M., Laureti, S., Nie, L., Freear, S., et al. (2020). Additive manufacturing of metamaterials: A review. *Additive Manufacturing, 36*, 101562.

Bauer, J., Kraus, J. A., Crook, C., Rimoli, J. J., & Valdevit, L. (2021). Tensegrity metamaterials: toward failure-resistant engineering systems through delocalized deformation. *Advanced Materials, 33*(10), 2005647.

de Castro Motta, J., Garanger, K., & Rimoli, J. J. (2024). Propagation of compression solitary waves on tensegrity-like lattices made of truncated octahedrons. *International Journal of Non-Linear Mechanics, 162*, 104716.

Fraternali, F., & de Castro Motta, J. (2023). Mechanics of superelastic tensegrity braces for timber frames equipped with buckling-restrained devices. *International Journal of Solids and Structures, 281*, 112414.

Fraternali, F., de Castro Motta, J., Germano, G., Babilio, E., & Amendola, A. (2024) Mechanical response of tensegrity-origami solar modules. *Applications in Engineering Science, 17*, 100174.

Fraternali, F., & Lee, H. (2024). Tensegrity wiki. https://tensegritywiki.com/.

Gan, X., Fei, G., Wang, J., Wang, Z., Lavorgna, M., & Xia, H. (2020). *Powder quality and electrical conductivity of selective laser sintered polymer composite components* (pp. 149–185). Woodhead Publishing. https://doi.org/10.1016/B978-0-12-819535-2.00006-5

Gao, X., Yu, N., & Li, J. (2020). *Influence of printing parameters and filament quality on structure and properties of polymer composite components used in the fields of automotive* (pp. 303–330). Woodhead Publishing. ISBN 9780128195352. https://doi.org/10.1016/B978-0-12-819535-2.00010-7

Han, D., Yang, C., Fang, N. X., & Lee, H. (2019). Rapid multi-material 3d printing with projection micro-stereolithography using dynamic fluidic control. *Additive Manufacturing, 27*, 606–615.

Harinarayana, V., & Shin, Y. C. (2021). Two-photon lithography for three-dimensional fabrication in micro/nanoscale regime: A comprehensive review. *Optics & Laser Technology, 142*(2), 107180. https://doi.org/10.1016/j.optlastec.2021.107180

Huang, J. (2020). A review of stereolithography: Processes and systems. *Processes, 8*(9), 1138. https://doi.org/10.3390/pr8091138

Hull, C. (1984). Apparatus for production of three-dimensional objects by stereolithography. US Patent 4,575,330.

Lee, H., Zhang, J., Jiang, H., & Fang, N. X. (2012). Prescribed pattern transformation in swelling gel tubes by elastic instability. *Physical Review Letters, 108*(21), 214304

Lee, H., Jang, Y., Choe, J.K., Lee, S., Song, H., Lee, J.P., Lone, N., & Kim, J. (2020). 3d-printed programmable tensegrity for soft robotics. *Science Robotics, 5*, eaay9024. https://doi.org/10.1126/scirobotics.aay9024.

Lewis, J. A. (2006). Direct ink writing of 3d functional materials. *Advanced Functional Materials, 16*(17), 2193–2204.

Liu, K., Wu, J., Paulino, G. H., & Jerry Qi, H. (2017). Programmable deployment of tensegrity structures by stimulus-responsive polymers. *Scientific Reports, 7*(1), 3511.

Pajunen, K., Celli, P., & Daraio, C. (2021). Prestrain-induced bandgap tuning in 3d-printed tensegrity-inspired lattice structures. *Extreme Mechanics Letters, 44*, 101236. https://doi.org/10.1016/j.eml.2021.101236

Pajunen, K., Johanns, P., Pal, R. K., Rimoli, J. J., & Daraio, C. (2019). Design and impact response of 3D-printable tensegrity-inspired structures. *Materials & Design, 182*, 107966. ISSN 0264-1275. https://doi.org/10.1016/j.matdes.2019.107966.

Placidi, L., de Castro Motta, J., & Fraternali, F. (2024). Bandgap structure of tensegrity mass–spring chains equipped with internal resonators. *Mechanics Research Communications, 137*, 104273.

Rimoli, J. J. (2018). A reduced-order model for the dynamic and post-buckling behavior of tensegrity structures. *Mechanics of Materials, 116*, 146–157.

Salahshoor, H., Pal, R. K., & Rimoli, J. J. (2018). Material symmetry phase transitions in three-dimensional tensegrity metamaterials. *Journal of the Mechanics and Physics of Solids, 119*, 382–399.

Song, X., Zhai, W., Huang, R., Fu, J., Fu, M. W., & Li, F. (2022). *Metal-based 3d-printed micro parts & structures* (pp. 448–461). Elsevier. https://doi.org/10.1016/B978-0-12-819726-4.00009-0.

Surjadi, J. U., Gao, L., Du, H., Li, X., Xiong, X., Fang, N. X., & Lu, Y. (2019). Mechanical metamaterials and their engineering applications. *Advanced Engineering Materials, 21*(3), 1800864. https://doi.org/10.1002/adem.201800864

Vangelatos, Z., Micheletti, A., Grigoropoulos, C. P., & Fraternali, F. (2020). Design and testing of bistable lattices with tensegrity architecture and nanoscale features fabricated by multiphoton lithography. *Nanomaterials, 10*(4). https://doi.org/10.3390/nano10040652.

Wen, L., Pan, F., & Ding, X. (2020). Tensegrity metamaterials for soft robotics. *Science Robotics, 5*, eabd9158. https://doi.org/10.1126/scirobotics.abd9158

Zeng, H., Mu, R., Huo, K., Zhao, H., Wang, K., & Wang, A. (2023). A novel 3d-printable tensegrity-inspired metamaterial enabling dynamic attenuation. *International Journal of Mechanics and Materials in Design*, 1–19. https://doi.org/10.1007/s10999-023-09656-7

Zheng, X., Deotte, J., Alonso, M. P., Farquar, G. R., Weisgraber, T. H., Gemberling, S., Lee, H., Fang, N., & Spadaccini, C. M. (2012). Design and optimization of a light-emitting diode projection micro-stereolithography three-dimensional manufacturing system. *Review of Scientific Instruments, 83*(12).